The Psychology of Time Perception

John Wearden

The Psychology of Time Perception

palgrave
macmillan

John Wearden
School of Psychology
Keele University
Keele, UK

ISBN 978-1-137-40882-2 ISBN 978-1-137-40883-9 (eBook)
DOI 10.1057/978-1-137-40883-9

Library of Congress Control Number: 2016938827

Printed on acid-free paper

This Palgrave Macmillan imprint is published by Springer Nature
The registered company is Macmillan Publishers Ltd. London

Preface

The composer Gustav Mahler is supposed to have said "A symphony must be like the world. It must contain everything." While this might be true of a Mahlerian symphony, it certainly is not true of this book on time perception. The volume you are reading does not contain everything that is known about time perception, or even everything that you might want to know. Its content is intended to provide the reader with an overview of some of the main trends in fairly recent work on the psychology of time, carried out over the last 30 or 40 years, although some historical issues are also discussed, as is a small amount of the philosophy of time. It is a personal selection of topics, albeit—I hope—a choice that is not too idiosyncratic. My aim is to provide the reader with an introduction to work in time perception which I believe to be interesting, important, and influential. An emphasis in the book, which reflects my own interests, is on ideas and theories underlying time perception research, rather than with just providing a summary of results, although a significant amount of experimental data is discussed, sometimes in considerable detail. The theories I mention have generally been simplified in the cause of exposition, and are all more complex than I am able to describe in this book, but the reader may consult the original articles cited in the text for details. I only hope the inventors of these theories agree that my discussion of their work has captured most of their essence, if not all the minutiae. Some topics that people might have wanted to read about, like rhythmical timing or time in music, are not included, as they have tended to develop separately from what I consider to be mainstream time perception, which has been strongly focussed on the perception of duration, the judgement of how long things seem to last. I have also omitted any discussion of the cognitive neuroscience of time perception. Although this topic occupies the attention of many researchers at the time of writing, thus far their efforts have resulted in few firm conclusions that can be easily communicated. Indeed, none of the topics treated in the book would be significantly illuminated in any way by the neuroscience of timing in its present state, at least in my view. Finally, the reader is warned that my own contribution to the field has been exaggerated here, but if you are not interested in your own work, why should anyone else be? In any case, the considerable effort of writing this book, which among other things has brought home to me how little I know about the subject, has surely earned me this privilege.

Keele, UK John Wearden

Acknowledgements

I am grateful to the American Psychological Association, Elsevier B.V., and Taylor & Francis for permission to reproduce many of the figures used in this book.

Contents

List of Figures

1

Overview

As mentioned in the preface, the aim of this book is to provide the reader with an introduction to what I believe to be the most important areas of time perception over the last 30 or 40 years. It is focussed mainly on the psychology of timing in humans, although Chap. 9 discusses animal timing. Animal timing has been an important precursor of ideas in the contemporary study of time perception in humans, mostly through the development of *scalar expectancy theory* (SET), initially proposed to explain animal timing, but in some ways the dominant theory of human timing until recently, and perhaps still. In general, however, the study of animal learning and behaviour plays a much smaller role in psychology today than in the past, and so presenting a very extensive account of animal timing early in this book seemed inappropriate, and likely to discourage many readers. To address this, I have placed the topic of animal timing in Chap. 9, where readers will find some of the most intellectually fascinating ideas in the whole of the field of time perception. Chapter 2 discusses a little of the philosophy of time, and some of the history of time perception research leading to the development of internal clock theories of human timing, which have been so influential. Chapter 3 discusses SET, the principal internal clock-based account of human timing,

J. Wearden, *The Psychology of Time Perception*,
DOI 10.1057/978-1-137-40883-9_1

itself a variant of an earlier theory proposed by Treisman (1963). An appendix to that chapter presents an elementary account of some of the mathematics of internal clock theory. SET provides not only a general account of human performance on many timing tasks, but also specific theoretical models, notably the performance on tasks involving temporal generalization and bisection, which have been widely used, particularly the latter. Theoretical accounts of generalization and bisection, whether based directly on SET or otherwise, are discussed in Chap. 4. In that chapter, mathematical exposition is kept to a minimum, largely as a result of my own lack of competence. A reader wishing to understand some of the basic mechanics of SET-based models, which have been particularly important in areas such as the developmental psychology of time, will find here what I hope are simple accounts of their operation. Chapter 5 considers the role of cognitive processes such as attention and executive function in time perception, but also discusses fairly recent work on emotion and time perception. Chapter 6 examines retrospective timing, or "timing without a timer," and passage of time judgements, including work from sociology and occupational psychology that is rarely mentioned in the mainstream time perception literature. In this chapter I also spend some time discussing the contents of Ornstein's (1969) book, *On the Experience of Time*, which is frequently cited, probably little read, and these days rather difficult to obtain. Chapter 7 deals with timing in children. I begin with a discussion of Piagetian time psychology, with an account of some of the material in Piaget's book *The Child's Conception of Time*. This is another classic, and once again, I suspect, one that is little read, although the peculiarities and obscurity of the work give modern readers good excuses to avoid it. Later researchers in the Piagetian tradition, on the other hand, have developed Piaget's ideas in a more digestible manner, and I discuss some of their work. The later part of this chapter discusses more recent research on timing in children, some of it inspired by SET. In Chap. 8 I review some of the work related to timing and ageing, including both laboratory studies and research on time experience in "real life," with some discussion of the vexed question of whether time seems to "go faster" as people age. Chapter 9, mentioned earlier, introduces some of the main findings from, and theoretical models applied to, animal timing research. Readers jaded by the hegemony of SET will find interesting competitors to it here. Chapter 10 is a kind of appendix, outlining some of the commonest methods used in time perception research. Readers unfamiliar with time perception studies will find outlines of some of the principal methods used here.

I should emphasize at the outset that this volume is not intended to be a "popular science" account of the psychology of time perception, but neither is it intended to be a highly technical work comprehensible only to specialists in the field. In this book, I have tried to introduce and explain the main trends in fairly recent time perception research in a way that should be accessible to anyone with a basic knowledge of psychology. I can only hope that I have succeeded with this aim, at least in part.

2

A Brief History of Time Perception

Philosophers and Time

The second part of Cardinal Newman's poem *The Dream of Gerontius*, set to music so vividly by Edward Elgar, describes the progress of the soul of Gerontius towards its judgement while beholding the sight of God. Newman tries to describe a situation without time, which the deceased Gerontius expresses like this:

> *How still it is!*
> *I hear no more the busy beat of time,*
> *No, nor my fluttering breath, nor struggling pulse;*
> *Nor does one moment differ from the next.*

Even with the augmentation of imagination that poetry sometimes elicits, this situation seems literally impossible for most people to conceive of (and for Newman as well: in the poem, events follow one another, occurring in a clear temporal sequence). The passage of time as a succession of events, or the feeling of passage of time during a persisting stimulus, seems such a necessary part of our everyday experience, that it is natural

J. Wearden, *The Psychology of Time Perception*,
DOI 10.1057/978-1-137-40883-9_2

to suggest it is in some way primordial and something that is essential for us to make sense of our impressions and the progress of our mental life.

Philosophers have long been fascinated by the nature of time and how we perceive it. Nichols (1891) provides a summary of many philosophical positions regarding time that were advanced through the late nineteenth century. One issue that divided philosophers and that has had practical consequences for the psychology of time is the question of whether the perception of time is innate or is based on some prior experience.

Kant, in his celebrated *Critique of Pure Reason* (1781/1900/2003), proposed that a sense of time existed a priori, along with the appreciation of space, as something innate and independent of experience. For example (p. 28)

> Time is not an empirical conception. For neither co-existence nor succession would be perceived by us, if the representation of time did not exist a priori…Time is a necessary representation, lying at the foundation of all our intuitions. With regard to phenomena in general, we cannot…represent them to ourselves as out of and unconnected with time, but we can quite well represent to ourselves time devoid of phenomena.

In addition, he argues (p. 29) that

> …the conception of motion….is possible only through and in the representation of time…

There is empirical research with children on the relations between time and change—and in particular, time and motion—and it is discussed in Chap. 7. As work reviewed there shows, it seems that judgements of time, or judgements of duration at least, are if anything *more* difficult to master than concepts like distance or size. Children also have considerable difficulty extracting the dimension of duration from other aspects of the situation presented to them. For example, if asked to judge which of two toy cars ran for the longer period of time, young children have difficulty dissociating the duration of the movement from the distance traversed. None of this suggests that time, at least in the form of conscious appreciation of duration, is an a priori phenomenon. However, as other research shows, particularly that reviewed in Chaps. 7 and 9, very young children who have not learned to speak and non-human animals can be trained to adjust to the temporal requirements of experimental situations, suggesting that sensitivity to temporal regularities in the environment is not dependent on the possession of language that might provide the cognitive concepts of time familiar to adults.

The Kantian concept that time is a "given" contrasts sharply with the position of Guyau (1890). Guyau's *La genèse de l'idée de temps* not only explores the question of how time representations develop, but also speculates—often with arguments that appear strikingly modern—about many aspects of time perception that would later come to occupy the attention of experimental psychologists. Guyau was certain that notions of time could not exist a priori, and instead developed with experience from a number of sources. The quotes from Guyau that follow are in my own translation.

> The genesis of the idea of time is ....empirical and derived. The idea of time, like that of space, is empirically the result of the adaptation of our activity and our desires… (p. 46).

The relation between the perception of time and space was also thought to be important, with concepts of time developing later than conceptions of space, according to Guyau, a position in accord with modern research, as mentioned above.

> It is movement in space which creates time in human consciousness. Without movement there is no time (p. 47)

But time concepts were also believed to be derived from

> …the notion of discrimination of differences, resemblances, number…and intensity (p. 22)

Guyau's ideas have been particularly influential among French-speaking psychologists, who have made substantial contributions to the study of time perception in general and, more specifically, to the developmental psychology of time, discussed in Chap. 7, where the issue of how temporal representations develop in young children has been of central importance. Guyau's ideas, however, also prefigure more recent discussions on the role of cognitive processes in time perception (Chap. 5), as well as ideas about "retrospective" time judgements (Chap. 6).

One question which has received particular attention from philosophers is what constitutes the *present moment*, or what James (1890) and others have called the *specious present*. In reading this text, one will likely have the impression of a sort of moving present. The line currently being read has some *present* quality, whereas earlier lines and paragraphs are located in the *past*, and there is something to come which is in the *future*. Two questions that have engaged both philosophers and psychologists concern the nature of this present experience and how long it actually lasts. Guyau regarded the present as a sort of small boundary between the past and future.

> The true present should be an indivisible instant, a moment of transition between the past and the future, a moment which cannot but be conceived of as infinitely small, being born and dying at the same time. (p. 30)

However, subjectively, at least to this writer, the present seems to have some short but non-zero duration, a position held by most philosophers, although there is little consensus as to the actual duration of the specious present. In philosophical circles, a particularly influential treatment of the specious present, and time consciousness in general, is derived from Husserl (see Mensch, 2014, for a brief introduction to Husserl's ideas).

Husserl used the "saddle" metaphor introduced by James (1890): a person rides on the specious present like someone saddled on a horse. The "temporal" saddle is not the edge of some infinitely sharp knife (contrasting with Guyau's position); it has some extent, if only a short one. Husserl proposed that the moving present is actually composed of three parts. One is the present per se, the "primal impression," what we actually experience at the present moment. The second is the recent past, what Husserl calls *retention*. So far, most people would agree that this accords nicely with common sense. However, Husserl then adds what he calls *protention*: some aspect of the future which is included as part of the present moment.

Philosophers more recently have dealt with similar issues, albeit in a rather different way. One of their most important questions concerns the *phenomenal continuum*, how we experience the passage of time as it unfolds. Dainton (2014) provides a good introduction to the often thorny problems uncovered by philosophers when talking about "temporally extended" experiences. When we experience an event that has several parts—Dainton's example is seeing and hearing a large rocket go off—there are visual and auditory aspects to the experience, plus a succession of events in time. Thus one question that arises is how "change and succession feature in our immediate experience" (Dainton, 2014, p. 101). Perhaps the most natural way of representing the event, and the one that Dainton himself advocates, is an "extensional" approach: the sights and sounds of the rocket are not momentary, but occur in a sort of temporal window, which is then replaced by another (perhaps partially overlapping) temporal window. However, another, perhaps less intuitively appealing approach, "the retentional model," has also been advanced. Here, the whole event is perceived in a "momentary episode of experiencing," which nevertheless may have multiple dimensions, giving it an extended subjective character, just as in some modern theories in physics, additional spatial dimensions are posited for elementary strings hypothesized to constitute the bases of matter. The full arguments are too complex for discussion here, but see Dainton (2014) for an introduction.

Another contemporary philosopher, Ian Phillips, has dealt extensively with questions regarding the philosophy of time, some of which interlock with the psychology of time perception. One peculiarity of duration noted by many philosophers is that the "sensation" (or *quale*) of time itself has temporal aspects. No-one thinks that the sensation of green is green, or that the feeling of a 200-g weight itself weighs 200 g, but the "feeling" of, say, 2 s extends in time; indeed, it may actually last for 2 s. Therefore, the qualia of time are peculiar in that they seem identical to the thing felt, a unique property according to Phillips (2014). He accounts for this by taking what he calls the "naïve" view (the term is not pejorative here), which states (Phillips, 2014, p. 142) "that our experience inherits the temporal structure of the events which are its contents." To return to our previous example, the quale of 2 s "inherits" the property of lasting for 2 s that the event observed possessed.

The naïve view is highly appealing to someone who knows nothing of the complexity of philosophical arguments, as it appears to avoid the seemingly counterintuitive problems that philosophers have invented about time, such as the "principle of simultaneous awareness," which states that if one is aware of duration or succession, one is aware of it at one moment, thus raising the question of how extended temporal experience is possible (a position reminiscent of the "retentional" view mentioned above). The naïve view does encounter some difficulties, however, particularly those dealing with temporal "illusions," where the physical duration of events, or even their order, can be misperceived. Phillips (2013, 2014) shows how many of them can be reconciled with the naïve view of time experience.

One of the most frequently cited—and to psychologists one of the strangest—philosophical positions regarding time is McTaggart's (1908) argument that time does not exist, initially advanced in an article dramatically entitled "The Unreality of Time." His reasoning, in simplified form, proceeds as follows. Time is about change, so any correct way of representing it must involve the possibility of change. McTaggart distinguishes two ways of describing events in time. One (called the "B-series") uses relations such as "earlier" and "later," as in "event X was earlier than event Y." This B-series itself cannot represent change, as the temporal relation of X and Y is fixed, so something else is needed. McTaggart thus proposes an "A-series," involving past, present, and future. Some event, therefore, passes from the future to the present, then to the past. This obviously represents change, so what then is the problem? The problem is that some event cannot simultaneously possess the properties of existing in the future, the present, and the past, so some way of distinguishing these possibilities needs to be developed. The way to do this is obvious if we have some concept of time: some event X is in the future at one time, the present at another time, and the past at yet another time. But the difficulty here is that we have assumed a concept of time to make the A-series non-contradictory (i.e. to avoid postulating, for example, that X is simultaneously in both the future and the past), yet the A-series is supposed to *provide* the very concept of time that is needed.

As McTaggart says, using the A-series

> involves a vicious circle. For it assumes the existence of time in order to account for the way in which moments are past, present and future. Time then must be pre-supposed to account for the A-series.

McTaggart suggests that these problems imply that time does not exist. Perhaps a less dramatic conclusion would be to say that the ways in which we talk about time have unexpected hidden depths. McTaggart's article has been described as advancing "clearly one of the greatest of philosophical arguments" (Geach, 1979), although, perhaps unsurprisingly, many commentators have disagreed with his conclusion. One such objector is Corish (2005), who questions aspects of the logic in McTaggart's distinction between the A- and B-series, and argues that aspects of the A-series can be included in the B-series, so the two are complementary, and simply reflect different kinds of temporal relations, meaning that the existence of both does not give rise to paradoxes. To use his example, the death of Queen Anne was, and always will be, before the death of McTaggart himself, so the B-series relation of these two events is fixed. However, depending on the date on which a statement is made, these events can both be in the past, one in the present and the other in the future, one in the present and the other in the past, or both in the future, and so the "permanent" B-series relation does not preclude the changing, "impermanent" A-series-type temporal relations with respect to other events. Corish's (2005) article also provides references to the complex arguments surrounding McTaggart's assertion of time's unreality, and the interested reader is referred there as a starting point.

The Early Years of Time Perception Studies

Experimental psychology, as it developed in Germany in the second half of the nineteenth century, expended much effort on the study of sensation, using simple stimuli in psychophysical experiments relating objective stimulus measures, such as intensity, to the sensations they evoked. Fechner's famous *Elements of Psychophysics* (1860) was a landmark work of this type. Along with those involving vision, hearing, and touch, studies of duration judgements were common—in fact, relatively more common than they are today, when aspects of time perception receive much less attention than, for example, visual perception. One basic idea was that a "sense of time" (*Zeitsinn*) existed, analogous to senses of vision and hearing, and that objective measurements of duration, obtained via mechanical and electrical clocks, could be related to judgements of duration made by individuals. These studies assumed that time could be perceived directly (the position that Phillips, 2013, calls "realism"), and that the duration of a stimulus or event was a perceivable property of the event, just as was its colour or associated sounds.

The most remarkable and one of the earliest products of these investigations was the book *Der Zeitsinn nach Versuchen* (literally "The time sense according to experiments," which Lejeune & Wearden, 2009, translate more idiomatically as "The experimental study of the time sense"), published by Karl von Vierordt in 1868. Lejeune and Wearden (2009) provide a fuller treatment of Vierordt's work than can be given here, and also relate some of his discoveries to modern research. Below I will refer to the book as simply *Der Zeitsinn*.

Vierordt was a physician who became a medical Professor at the University of Tubingen (and later rector of the university). His early work was what we would now call experimental physiology, and one of his main concerns was to make inner physiological processes outwardly measurable. He was the first to propose the principle behind the modern pressure cuff method of measuring blood pressure (as well as inventing a complex device for measuring it), and developed other ingenious methods for measuring lung capacity and postural sway. Given this emphasis on making internal physiological processes visible, it is perhaps no surprise that he became involved in psychophysics, the ultimate challenge of this sort.

Vierordt and his students were involved in experiments on the measurement of sensation as early as the 1850s, and he later turned to experiments on judgements of duration. The thesis of Vierordt's student Höring in 1864 was among the earliest reports of an experimental study of time perception, perhaps even the very first. *Der Zeitsinn* reports a large number of experiments on duration perception, many carried out by Vierordt himself as experimental subject, or by his assistants. In many, although not all, of his studies, Vierordt employed various reproduction techniques, using methods discussed in Chap. 10. One of his most striking findings later became known as *Vierordt's Law*. Suppose that a person reproduced different intervals within some range. Vierordt's Law was the finding that the shortest intervals in the range tended to be reproduced as longer than they really were, whereas the longer intervals were reproduced as shorter than in reality. Somewhere in the middle was an "indifference point," where the time interval was reproduced accurately.

Lejeune and Wearden (2009) present many examples of Vierordt's Law from data provided in *Der Zeitsinn*, and I discuss other data conforming to Vierordt's Law in Wearden (2003). When short time intervals are reproduced, conformity to Vierordt's Law is a common finding—see, for example, Jazayeri and Shadlen (2010)—although it is not universally found, particular when data from individual persons are examined (see Woodrow, 1930, for example). But Vierordt did more than just reproduction experiments. Some of his studies involved the perception of differences in the frequency of metronome ticks (a sort of rhythm perception); he also studied the verbal estimation of the duration of very long intervals, and was one of the first to note that durations of auditory and visual stimuli were perceived differently.

Vierordt's "indifference point" captured the imagination of psychologists who immediately followed him, as well as those much later. Perhaps, people thought, there was something special about the intervals that were reproduced accurately, even to the extent that the indifference point represented some kind of fundamental unit of time perception. However, as Lejeune and Wearden (2009) show, using Vierordt's own data and those of others, the indifference point was certainly not a fixed absolute value, and instead seemed to depend on the range of intervals used. This led to the idea that the indifference point might be related to some central tendency of all the intervals used in the study, such as the mean or the median. Hollingworth (1910), in an article which prefigures the later adaptation-level theory of Helson (1964), proposed exactly this sort of explanation. Lejeune and Wearden (2009) present examples where this idea works well, as the indifference point seems to closely track the mean of the intervals used. However, there are other cases in which the indifference point is nowhere near any kind of mean, and cases where no central tendency can be formed (e.g. in situations where only one interval is judged, or even a single trial is used). Although different instances of conformity to Vierordt's Law may resemble one another, it seems that no common explanation can be found. However, interest in Vierordt's Law has persisted; for some recent ideas, see Gu and Meck (2011) and Bausenhart, Dyjas, and Ulrich (2014).

Wilhelm Wundt was a medical student at Tubingen when Vierordt was a professor there, although there seems to be no record of their meeting, let alone any suggestion that Vierordt's psychophysical work had any influence on Wundt. Nevertheless, Wundt's famous laboratory in Leipzig also carried out time perception studies, as did other German laboratories. Nichols (1891) provides a review of their findings, and the location of the indifference point was an important focus of interest. A value of around 0.75 s was the preference of many researchers, although as Lejeune and Wearden (2009) show, Vierordt's own work rarely found the indifference point at this value. In fact, this value of just less than 1 s later assumed an almost mystical significance. Some thought it defined the duration of the "specious present" discussed above: the time period seemingly "now," which is neither clearly in the past nor the future, an empirical measure of Husserl's "primal impression." A related idea proposes that the 0.75-s interval represents some sort of "window of integration" where events actually spaced in time seem unified. Fraisse (1967), following this idea, gives the example of the cry of a bird: even though this cry is extended in time, it is perceived as having some sort of unity, an idea reminiscent of the "extensional" approach to temporal experience advocated by Dainton (2014). We can see that Vierordt's work, reported in *Der Zeitsinn*, although rarely examined in detail these days, casts a very long historical shadow.

Another issue studied in the early days of experimental psychology which has had modern resonance is the question of whether sensory judgements, and judgements of duration in particular, obey *Weber's Law*. Suppose that we present people with two stimuli, and they must decide simply whether the stimuli are different—for example, whether they have the same duration. As the differences are varied across trials, there will be some differences that can be discriminated, whereas others will be too small to judge correctly. What is the smallest difference that can be reliably detected a certain proportion of the time, such as in 75 % of presentations? Let us call this the "just-noticeable difference" (JND). On what does this depend? Weber's Law asserted that the JND was a constant proportion of the magnitude of the stimulus to be judged, such that the JND divided by the stimulus magnitude was a constant value—the *Weber fraction*—and that having a constant Weber fraction as the stimulus magnitude changed was conformity to Weber's Law. Weber's Law is easy to link to sensory judgements both in the laboratory and in real life. If two noises presented are quiet, then a small difference between them may be easily detectable, but if they are much louder, the difference may not be large enough to be perceived. If you have a slim and light friend who weighs 50 kg, it may be very obvious if they gain a small amount of weight, say 2 or 3 kg, but the same weight gain would be more difficult to detect in a much heavier person, one weighing 100 kg, for example.

The Weber fraction can be regarded as a measure of the fineness of sensory discrimination, regardless of whether judgements in general conform closely to Weber's Law. If the Weber fraction is small, then small differences between the stimuli are detectable, and the discrimination is more sensitive than if the Weber fraction is larger, and the relative difference between two stimulus magnitudes must be much larger to be detectable. As will be seen later (in Chaps. 3 and 7), conformity to Weber's Law is a hallmark of certain recent theories of time discrimination, and the use of Weber-fraction-like measures to compare the sensitivity of time judgements (e.g. between groups of children of different ages) is common place.

Chemical Clocks

The study of time perception is obviously hampered by the inability to identify any organ which is obviously responsible for it. Whereas vision or hearing science can begin with the physical properties of the sense organ subserving visual or auditory perception, something which has aided both the psychology and physiology of vision and audition, a person interested in time perception has no such clear starting point. Nevertheless, an extremely influential idea has been that there is a type of "organ" for time, in the form of an *internal clock* of sorts.

In the English-speaking world, the name of Hudson Hoagland is usually associated with the concept that humans possess some sort of *chemical clock*, a notion which was a precursor of internal clock models of timing developed in the 1960s, and which are discussed later. However, historical precedence in this area actually belongs to Marcel François, a pupil of the famous French psychologist Piéron, who in 1927 published a study in which he used diathermy (the passage of high-frequency electric current through the body) to induce bodily heating, and observed its effects on time judgements.

The basic idea behind the study of body temperature and time estimation is that, by laws of physical chemistry, all chemical reactions occur more rapidly when heated, and thus any chemical reaction underlying a chemical clock should proceed more quickly when the temperature is elevated and more slowly when it is lowered, producing some effect on time judgements. François (1927) performed a number of studies, some of which are not fully reported in his article, but most of the data come from situations in which individuals were asked to tap at a certain rhythm (e.g. 3 taps per second). When the temperature was elevated, rates of tapping generally increased, that is, the time intervals between taps were shortened.

Hoagland's own work (1933, 1935, 1951), which these days is much better known than that of François, was initially inspired by a real-life incident (recounted in Hoagland, 1935). Hoagland's wife was ill with influenza, which raised her body temperature. During the period he was nursing her, he went out for a time, and when he returned, Mrs. Hoagland thought that a long period had passed, when it fact the real interval was much shorter. Why did she think this? Hoagland reasoned that the illness had raised her body temperature, thus speeding up the activity of the "chemical clock," which "ticked" faster in her fevered state. Thus a greater number of "ticks" accumulated than would normally be the case in any fixed time interval, leading her to judge the time intervals she experienced as abnormally long. Hoagland then collected systematic data on his wife as her body temperature increased and decreased, asking her to count to 60 at a rate that she perceived as one value per second. Figure 2.1 shows the results: the higher her fever, the faster she counted, meaning that her production of the 1-s time unit was systematically

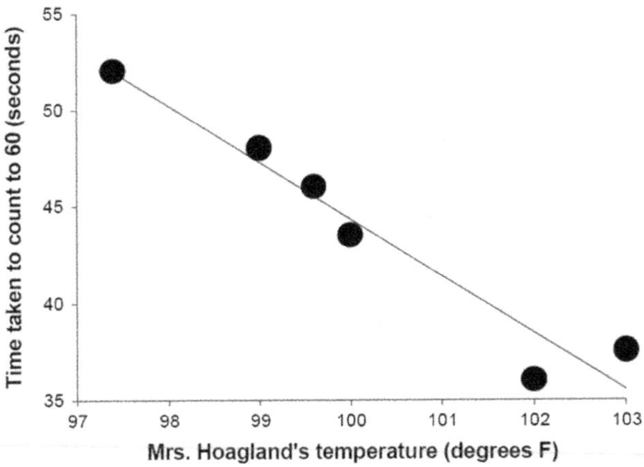

Fig. 2.1 Time taken to count to 60 at a rate judged to be 1 count per second, plotted against body temperature in degrees Fahrenheit. Data from Mrs. Hoagland taken from Hoagland (1935)

shortened with the rising temperature. This mirrors the result obtained in François's study, and is exactly the result expected with a "speeded-up" clock: external stimuli and events are perceived to last longer than normal (as more ticks are accumulated than is normally the case), whereas intervals produced are shortened (as the number of ticks associated with a given interval, such as 1 s, is accumulated in a shorter time as the clock ticks faster).

The work of François and Hoagland gave rise to what must be one of the most bizarre fields of study ever undertaken by serious experimental psychologists: experiments on the effect of body temperature on time judgements. Changing a person's core body temperature is difficult, and a variety of methods have been used, including superheated rooms (Bell & Provins, 1963), heated helmets (Hancock, 1993), natural fevers (echoing Mrs. Hoagland's experience, Alderson, 1974), and diurnal fluctuations in body temperature (Pfaff, 1968). Wearden and Penton-Voak (1995) reviewed the majority of published studies at the time. Many of the articles had made no clear distinction between prospective and retrospective timing, discussed in Chap. 6, as the research was conducted before the difference was commonly used, and so the authors examined only studies which were prospective in nature, in which participants knew in advance that timing was an important part of the task they were performing. They also converted data from production and estimation measures into a common metric of the rate of subjective time, revealing a clear relationship between changes in body temperature compared to a normal value, and changes in the rate of subjective time: increases above normal body temperature increased the rate of subjective time, and decreases in body temperature (which were much rarer, coming from only two studies: Baddeley, 1966, and Bell, 1975) decreased the rate of subjective time.

In general terms, then, data from body temperature studies seemed to confirm the notion of an underlying chemical or biological clock which was sensitive to body temperature or some correlate of body temperature (Wearden and Penton-Voak discuss arousal as a possibility). However, the nature of the internal clock, and the exact mechanism by which its readings are translated into behaviour, is not illuminated by these types of studies.

Although Hoagland and François are regarded as the "founding fathers" of internal clock theory, neither made any attempt to construct what we would regard as a psychological model of timing, still less a model having any mechanistic detail. Their preoccupation was, in fact, very different from that of later psychologists, in that they were interested in using behavioural data to identify the type of chemical reaction underlying the "chemical clock," via the mechanisms of physical chemistry.

The relation between temperature and the speed of a chemical reaction is described by equations derived by Van 't Hoff (used by François) and by Arrhenius (used by Hoagland). These equations can be fitted to data to derive an "activation constant," which is characteristic of a particular chemical reaction, and thus knowledge of the constant provides a clue as to the nature of the reaction, and experiments can be compared to determine whether the "constant" value is consistent across studies. Hoagland (1935, 1951) provides several examples of data showing apparent consistent activation constant values, thus concluding that "these results definitely imply the existence of a unitary chemical process serving as the basis for the subjective time scale…." (Hoagland, 1935, pp. 111–112).

Bell (1965,1966, 1975) was a consistent critic of the chemical clock idea, even though (as Wearden & Penton-Voak, 1995, show) his data often offered full or partial support for the idea of a temperature-sensitive internal timer of some sort (see also Green & Simpson, 1977). His objections to Hoagland's idea were twofold. Firstly, effects of body temperature on timing had to be consistent across different individuals: no-one should show, for example, an increase in body temperature and a decrease or no change in the rate of subjective time. Secondly, the "activation constant" had to be both plausible in value and consistent across studies and individuals. This latter requirement was certainly violated in a number of Bell's studies. For example, Bell (1966) found marked differences in the activation constant across individuals (a nearly 15-fold difference between people). However, in his study, there was a significant increase in tapping rate with increased body temperature (as in François' work and that of Hoagland), so his behavioural data actually supported the idea of a temperature-sensitive internal timer. As to Bell's first objection, he seems to have been applying a harsh criterion which would be met in few, if any, experiments in cognitive psychology. What established effect in memory and perception would stand if every single participant were required to react in a near-identical way to the experimental procedure?

Overall, therefore, the importance of Hoagland's and François' work largely seems to be the inspiration it provided for internal clock theories developed in the 1960s, and not the findings resulting from their ambitious attempts to measure a constant of physical chemistry from results of psychological experiments, something most would regard as a highly risky venture. I turn next to the development of more clearly specified internal clock models.

The Dawn of Internal Clock Models: Creelman and Treisman

The early 1960s saw the publication of two articles which laid the foundation for what was later identified as internal clock theory; these were the works of Creelman (1962) and Treisman (1963). Creelman's experimental procedure involved highly practised participants discriminating what were usually small differences in the duration of successive auditory stimuli presented through headphones. The task used is not described clearly in the article, but seems to have involved deciding whether the first or second stimulus was longer in duration. Across experiments, a number of variables were manipulated, such as the range of duration of the stimuli (although all were less than 1 s) and the degree of difference between them. Of most interest historically is Creelman's approach to explaining his data in terms of a "theory of time discrimination" (see his pages 590–592). He proposed a clock-like mechanism involving a source of pulses, along with a mechanism for counting these pulses, so that a different number of pulses would be counted when two stimuli of different durations were presented, and these numbers could then be entered into a decision mechanism. This provides a basic pacemaker-counter-decision clock, but Creelman also noted that other considerations might influence a person's time judgement in practice. For example, the participant would have to register the starting and stopping points of each stimulus, and would also need to remember the number of pulses accumulated for the first stimulus while timing the second. Creelman developed equations which took into account memory factors and the possibility that detection of the start and end of the stimuli (which were presented against a background of noise) would not be constant from one trial to the next. In general, there was a reasonable level of agreement between his model and the obtained data.

Creelman's approach was the first properly developed model of an "internal clock" process, and suggested that such a model involving an accumulation of "pulses" or "ticks" provided by a pacemaker could provide a viable starting point for the development of quantitative accounts of timing behaviour, at least in simple cases.

If the work by Creelman set the scene for the development of clock models of timing, an article by Treisman (1963) then provided a much more detailed account of how an internal clock, accompanied by some additional mechanisms, might provide a general account of performance on timing tasks. Treisman (2013) has recounted how his model came to be developed. Ironically, he initially had no interest in time perception, and was instead working on a phenomenon in visual perception, that of "electric phosphenes," a situation in which the passage of a weak electric current through the retina (via a metal band around the head) provokes the perception of a flash of light. He found that exposure to a cue stimulus (either a flash of light or a tone) changed the threshold for detecting the phosphene in a time-related manner: the longer the time elapsed since the cue, the lower the threshold for detecting the phosphene when an electric current was subsequently applied. He described this as a "warning effect," and the development of what would become his model of the internal clock was motivated by a desire to understand how people's judgements could be sensitive to time relations between stimuli, in this case between the cue and the electrical stimulus.

A sketch of the Treisman (1963) model is provided in Fig. 2.2. In addition to an internal clock, Treisman provided what he later (2013) described as the "clockwork" to enable the translation of time representations generated by the clock into time judgements in a number of different tasks.

The model consists of three parts. The first is the pacemaker-counter internal clock itself. A pacemaker, in which the pulse rate is determined

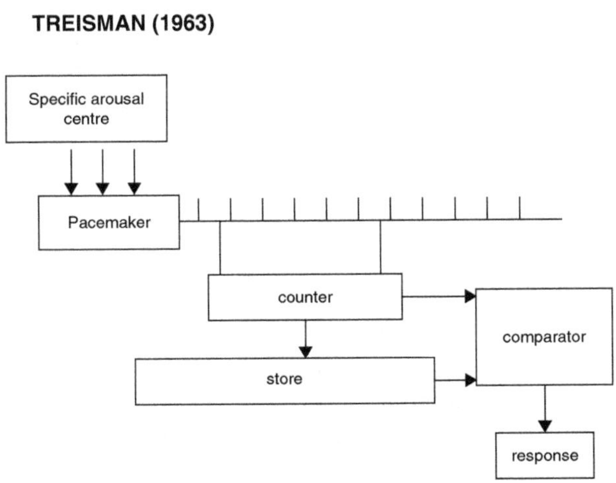

Fig. 2.2 Sketch of the internal clock model proposed by Treisman (1963). A pacemaker, the rate of which is affected by the activation level of a specific arousal centre, produces regular pulses which are counted in a counter mechanism, and which can be stored in a longer-term store. The contents of the counter and values retrieved from the store can be compared in a comparator mechanism, and the results of this comparison used to generate a behavioural response

by the current arousal level, produces regular pulses which flow along a pathway and which can be gated to a counter. So, for example, when an event begins, pulses begin to accumulate in the counter, and when the event ends, the accumulation stops. Output from the counter can also be stored—for example, the number of counts corresponding to a previously presented stimulus. Finally, the contents of the counter and a "readout" from the store are both entered into a comparator mechanism, where some decision process induces them to produce an overt response. It is easy to see how such a mechanism can perform a discrimination task like that described by Creelman (1962). With the initiation of the first of the two presented stimuli, the counts begin to accumulate; when the stimulus ends, the accumulation stops and the count number is passed into the store. The counter then resets and accumulates ticks for the second stimulus. Finally, the number of counts for the first stimulus (from the store) and the second stimulus (from the counter) are entered into the comparator, and a decision can be made as to which stimulus is longer.

In his 1963 article, Treisman reported seven experiments using reproduction, production, and discrimination methods, such as the presentation of two consecutive stimuli, with participants instructed to decide which was longer in duration, a task similar to that used by Creelman (1962). The experiments all used a small participant population, who were nevertheless exposed to many hundreds of trials of the experimental procedures. Durations ranged from 0.25 to 9 s, although a range of 0.5 to 3 s was common to a number of studies. The data were complex, and the interested reader is referred to the original publication for details. I will mention only two results here. One, which involved the production and reproduction tasks, revealed that Vierordt's Law, discussed above, was commonly (although not universally) obeyed. In general, participants produced and reproduced the shortest intervals in the duration sets as longer than they actually were, and the longest intervals as shorter than their real duration. Another finding was that intervals reproduced tended to become longer as the experimental session proceeded.

Treisman then attempted quantitative modelling of his data using the account outlined in Fig. 2.2. Once again, the details of how the model was fitted to data are too complicated for full discussion here, but one striking feature is how the simple form of the model needed the addition of an "error" or "distortion" to account for experimental data. To simplify somewhat, suppose that a person reproduces an interval of *S* s. The target interval *S* is presented, and the clock runs while this stimulus is on, resulting in the accumulation of *n* clock ticks, which are stored. The reproduction phase, *R*, then begins, and the comparator mechanism counts ticks until *n* have been accumulated, and reproduction *R* is then terminated. Thus, in theory, *R* should equal *S* for whatever value *S* assumes, whereas in practice, Vierordt-like deviations are found. Treisman accounted for these deviations by introducing what in effect appear to be distortions in the retrieval of *n* (which represents *S*): see his page 20 for details. With these added factors, correspondence between the data and the model was good, but possibly only at the price of adding parameters to the model which may be hard to justify in psychological terms.

One idea of Treisman's which has been influential in subsequent and even current research, however, is the proposal that the pacemaker of the internal clock was sensitive to arousal, "ticking" faster when arousal was high and more slowly when arousal was lower. Treisman used this mechanism to account for the lengthening effect found in successive blocks of reproduction in his experiment. For example, suppose that some target duration *S* in a reproduction experiment is represented by *n* clock ticks. As arousal diminishes as the experiment progresses and the pacemaker slows, a longer real time is needed to accumulate these *n* ticks, so reproduced intervals increase in duration. The basic premise of an arousal-sensitive internal clock has been extremely influential; see articles by Penton-Voak, Edwards, Percival, and Wearden (1996) and Wearden (2008a), as well as research on emotion and timing discussed in Chap. 5.

As Treisman (2013) notes, even though his model elaborated and developed ideas similar to those of Creelman, and provided a starting point for quantitative modelling of timing behaviour, it attracted little interest, and it is mysteriously absent even from some recent books on timing. For example, an award-winning popular account of time perception by Hammond (2012) makes no mention of any of Treisman's work. Even in the scientific literature, it had received only a few tens of citations 20 years or more after publication (Wearden, 2013a). The reasons for this neglect are unclear. The study of time perception in humans seemed to fall into eclipse in the later 1960s and 1970s, the model was complex and perhaps difficult to apply to experimental results. The fact that Treisman's model was tested only on a restricted body of data, with a lack of any large and important data set to which the ideas could be applied, was perhaps part of the problem.

No such restriction applied to the very similar proposition advanced in the form of Gibbon, Church, and Meck's *scalar expectancy theory* (SET: 1984), developed from earlier work by Gibbon (1977) and Church and Gibbon (1982). SET was found to be applicable to a considerable body of data from timing in animals, much of it collected under rigorous experimental conditions, and thus highly reliable. Further details regarding the application of SET to animal timing are provided in Chap. 9. The next chapter concentrates on the application of SET to timing in humans, interest in which developed towards the end of the 1980s (e.g., Wearden & McShane, 1988). In 1991, three articles from two different laboratories (Allan & Gibbon, 1991; Wearden, 1991a, 1991b) used SET explicitly for the first time to account for human performance on timing tasks.

Summary

The concept of time and how it is perceived has long fascinated philosophers, and one controversial issue has been whether the perception of time is innate (a position taken by Kant) or develops with experience (Guyau). Another problem exercising philosophers is how events extended in time can be perceived, and what constitutes the experience of the "present moment." Early research in experimental psychology often studied time perception, with a particularly important body of work, and one which continues to attract interest today, derived from Vierordt (1868). Specifically, explanations of *Vierordt's Law* (the finding that short durations tend to be judged as longer than they really are, whereas longer durations are judged as shorter than their real value) continue to be advanced in modern work. Although there is no obvious organ for time perception, the idea of an internal clock-like device developed in the 1920s, largely as a result of speculation about the chemical basis of the effect of changes in body temperature on time judgements. In the 1960s, quantitative psychological theories of time perception were developed by Creelman (1962) and Treisman (1963), the latter of whom embedded a pacemaker-accumulator clock in a more general psychological mechanism involving both memory stores and decision processes, and paved the way for the development of SET.

3

SET and Human Timing

Basic Principles

SET was developed in several different forms. One of these was an information-processing account, shown in Fig. 3.1, which bears a striking resemblance to Treisman's model from 1963. Stimuli and other events are timed by a pacemaker-accumulator internal clock similar to that proposed by Creelman (1962) and Treisman (1963), but the SET model involved more than just an internal clock. Two memory stores were proposed. One was a working memory, intended to retain time representations temporarily, and reflecting, more or less faithfully, the contents of the accumulator. In fact, some versions of SET combine the accumulator and working memory. There is, however, another memory store, one which retains references or standards which are used for a number of trials or a whole experimental session. SET's treatment of data from animal experiments is discussed in greater detail in Chap. 9, but the difference between the memory stores proposed by SET can be nicely illustrated by a simple procedure with animals.

On a fixed-interval (FI) schedule of reinforcement, the first response (usually a lever-press or key-peck) occurring *t* s or more from the previous

J. Wearden, *The Psychology of Time Perception*,
DOI 10.1057/978-1-137-40883-9_3

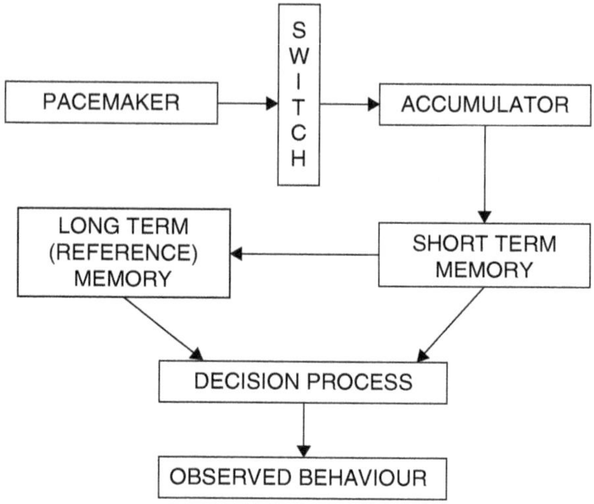

Fig. 3.1 Diagram of SET. The pacemaker is connected to the accumulator via a switch. Accumulator contents are transferred to a working memory store, and if the event timed constitutes a "standard" duration, are then transferred to the reference memory. To generate behaviour, the contents of working memory are compared with a sample taken from reference memory, and a decision process operates on these two time representations to produce an observed response

food delivery is rewarded, and this resets the interval. In effect, this means that the animal receives food deliveries with almost completely regular temporal spacing. In the case of an FI schedule, the distinction between working memory and reference memory is easily made. The reference memory contains a representation of the time *t*—30, 60 s, or whatever the FI value is—the time between availability and, usually, delivery of food. This value is constant for the whole experimental condition. In contrast, working memory reflects time elapsed in the interval and is changing continuously. The lengthy process of training animals on timing tasks, requiring many hours of exposure to constant experimental conditions, naturally encouraged the view that reference memory was more "long-term" or "permanent" than the constantly changing working memory.

In the case of timing in humans, where "training" is much shorter or sometimes non-existent, reference memory need not suppose any long-term or permanent character, although reference memory is commonly thought to contain some sort of standard, valid for a number of trials.

The final stage of the SET system is a decision process. Treisman (1963) recognised the need for some sort of decision mechanism, in the form of his comparator process. With SET, however, decision processes assume central importance. The decision process used depends on the timing task required of a person or animal. Individuals may compare a just-presented duration with a single standard previously presented (temporal generalization), and decide whether they are the same. Alternatively, a long and a short standard may have been previously presented, and the participant's task is to decide whether a just-presented duration is closer to the short or the long standard (temporal bisection). Decision processes play an important role in SET. In fact, they may perform the majority of the "work" of predicting behaviour or fitting data, and their role in some formal theoretical models is discussed in Chap. 4. The fact that different decision processes can be posited for different tasks affords the SET system a high degree of flexibility, and some authors (e.g. Staddon & Higa, 1999) have complained that this causes SET to be close to unfalsifiable. This problem has been recognised (Wearden, 1999) and will be discussed later; intuitively, however, it is surely not unreasonable to suppose that when people have to make different sorts of judgements about duration, they use different decision processes.

SET derives its name from the *scalar properties* of time that it presupposes. The term "scalar" is derived from the word "scale", and there are different ways of illustrating the basic idea. Suppose a person is timing different intervals *t*, 2*t*, 3*t*, and so on. One way of conceiving scalar timing is to suppose that the timing of these different intervals is *relatively* the same, just expressed on different scales. The scalar properties of timing are perhaps easier to illustrate visually than to describe verbally, and Figs. 3.2, 3.3, 3.4, and 3.5 show examples of data conforming to scalar timing.

Scalar timing requires that time representations have two properties. The first is *mean accuracy*, the requirement that an organism's "estimate" of some real time, *t*, is on average equal to *t*. So, for example,

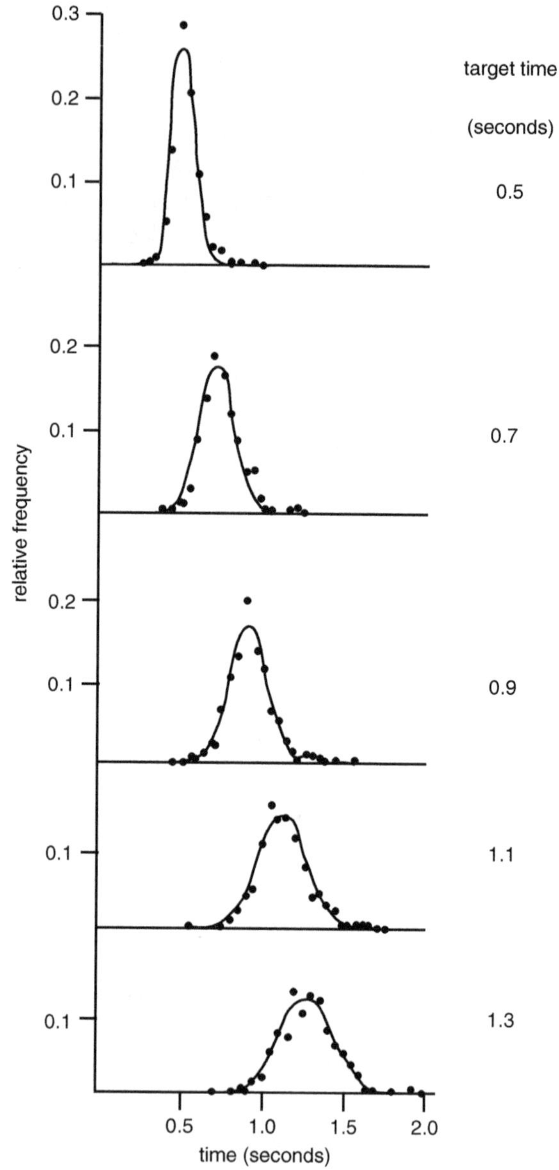

Fig. 3.2 Data from Wearden and McShane (1988). Four participants repeatedly produced time intervals ranging from 0.5 to 1.3 s, and received feedback after their responses. The data shown are the relative frequencies of times produced plotted against their duration, and are shown separately for the different time requirements. The curves shown are best-fitting Gaussian functions

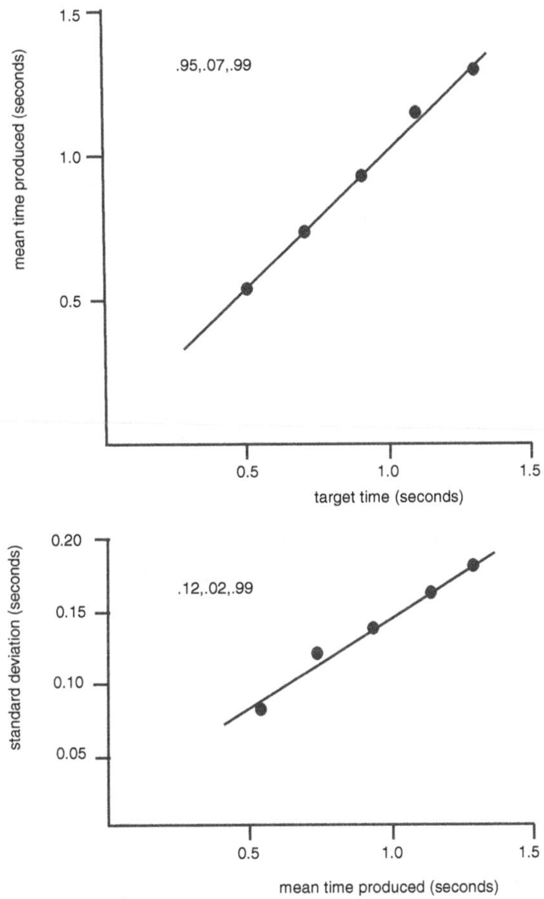

Fig. 3.3 Data from Wearden and McShane (1988). *Upper panel*: Mean times produced plotted against target time. The *line* shown is the best-fitting regression line, and the numbers in the panel indicate slope, intercept, and r^2 values for the regression. *Lower panel*: Standard deviations of times produced plotted against their means. Once again, the regression line and slope, intercept, and r^2 values are given in the panel

if the experimental requirement is to produce some time *t*, the average time produced should track *t* perfectly as the time requirement varies. Figure 3.2 shows data from an early article on scalar timing in humans, that of Wearden and McShane (1988). The task was to produce target time intervals ranging from 0.5 to 1.3 s, all values too short to make

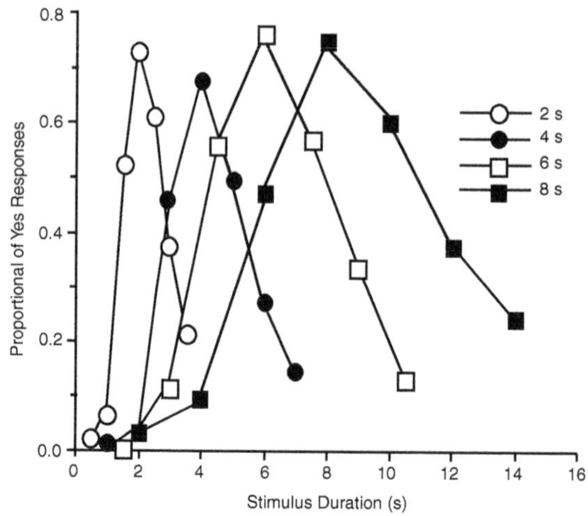

Fig. 3.4 Temporal generalization gradients from Wearden, Denovan, Fakhri, and Haworth. (1997). The standard durations were 2, 4, 6, and 8 s, and the data shown are the proportion of YES responses (judgements that a comparison duration was of the same duration as the standard), plotted against comparison stimulus duration

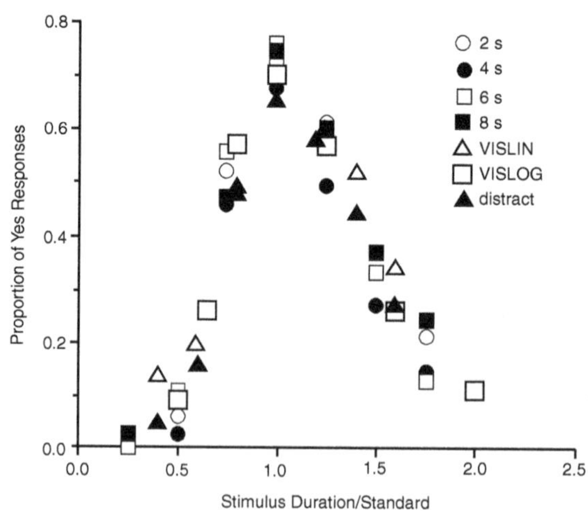

Fig. 3.5 Superimposition data from Wearden, Denovan, Fakhri, and Haworth (1997). Temporal generalization gradients from the conditions shown in Fig. 3.4 plus some others are plotted on a relative scale. For this, the duration of each comparison stimulus was divided by the standard in force for that condition

counting useful. One way of presenting the data is in the form of frequency distributions of the times produced, and Gaussian curves were fitted to them, as shown in Fig. 3.2. The peak of the Gaussian curve reflects its mean, and in the data the location of this peak was within 0.02 s of the target time as this varied, thus showing almost perfect mean accuracy. The upper panel of Fig. 3.3 shows the mean times produced in Wearden and McShane's study plotted against target time. Obviously, the means vary as a straight line function of the target time requirement, with a slope close to 1.0.

On some tasks, however, the timing requirement may necessitate an individual responding at some fraction of the target time. In these cases (see Wearden & Lejeune, 2008 for examples), the response should track some *proportion* of the target time as this varies. In other words, the mean response time should be a linear function of the time requirement as the time requirement changes.

The second property of scalar timing is the *scalar property of variance*, illustrated in Fig. 3.2 by the fact that the dispersion of responses around the mean varies as the interval timed varies. One way of showing conformity to the scalar property is to plot the standard deviation of response measures against their mean. For the scalar property of variance to apply, the standard deviation of response measures should be a linear function of their mean, and the lower panel of Fig. 3.3 shows data from Wearden and McShane (1988) where this is done. Another way is to construct a *coefficient of variation* (standard deviation/mean), which should remain constant as the interval timed varies. This coefficient of variation (cv) is a measure of timing sensitivity: the smaller it is, the closer response measures cluster around the mean, indicating greater sensitivity of timing. The cv is, in fact, similar to the classical Weber fraction (mentioned in Chap. 2), so a constant cv implies a constant Weber fraction, thus timing with the same sensitivity regardless of the interval timed.

A particularly powerful demonstration of scalar timing is *superimposition* (called *superposition* in the U.S. literature). This is the requirement that timing measures superimpose when plotted on the same relative scale, and can be illustrated using data from an experiment by Wearden, Denovan, Fakhri, and Haworth (1997), which used a temporal generalization method. Individuals initially received standard durations in the form of tones lasting either 2, 4, 6, or 8 s, with counting prevented by a secondary task. After a particular standard had been presented, comparison durations—some shorter than the standard, some longer, and some equal to it—were presented. The participant had to decide whether each comparison duration was the standard, indicated by a YES or NO response. Figure 3.4 shows the proportion of YES responses (identification of a comparison as the standard) plotted against comparison duration.

If we look first at the location of the peak of YES responses for each standard duration, we see that it tracks the standard value perfectly. When the standard was 2 s the peak was at 2 s; when it was at 4 s, the peak was at 4 s, and so on. This, of course, is *mean accuracy*, the first requirement of scalar timing. Next, consider the width of each curve. This obviously changes as the standard varies, becoming systematically wider as the standard lengthens. The reader may also gain the impression that the width grows proportionally to the standard, being twice as wide at 4 s than 2 s, for example. Figure 3.5 shows the same data plotted on a relative scale, where each comparison duration is expressed as a *proportion* of the standard in force for the condition, such that all the comparison durations for the 2-s standard are divided by 2, all those for the 4-s standard by 4, and so on. When this is done, the data from the different standard durations superimpose almost perfectly, a graphic example of the *scalar property of variance*, showing that timing is *relatively* the same at all the intervals employed. So, although the dispersion of responses around the 2-s standard is smaller than for the 8-s standard, the dispersion is always the same proportion of the standard value.

An intuitive means of grasping this scalar variance property is to think of timing in relative terms, as proportions, for example. A person might easily distinguish time intervals of 1 and 2 s. Distinguishing 11 from 12 s is much more difficult, and distinguishing 111 and 112 s almost impossible, so the same *absolute* difference between two time intervals is clearly not the basis for their discrimination. Scalar timing would contend that two intervals are equally easy to discriminate when they differ by the same *proportion*, not the same absolute amount; therefore, scalar timing embodies the basic principle of Weber's Law, discussed in Chap. 2, that the "just noticeable difference" between two quantities is a proportion (the Weber fraction) of their magnitude, and thus incorporates Weber's Law into the domain of time.

Although SET supposes that underlying representations of time possess the two scalar properties, mean accuracy and scalar variance, this does not necessarily mean that timed responses observed in experiments *directly* manifest these properties. They may do so, as in Wearden and McShane's data (Figs. 3.2 and 3.3), but in other cases they may not, as other processes such as those involving memory and decision mechanisms play a role in determining behaviour. The essential feature of SET, however, is that internal time representations, which are the "raw material" for overt behaviour, have the two scalar properties.

The temporal generalization procedure discussed above provides a good illustration of the whole SET system in action. Initially, a few examples of the standard are presented. When the standard stimulus begins, the switch connecting the pacemaker to the accumulator closes, and pulses flow until the stimulus ends, at which point the switch opens, cutting the connection. The number of pulses in the accumulator, representing the duration of the standard, is then transferred to working memory, and because the stimulus has been identified as a standard valid for a number of trials, it is transferred to reference memory. After a few standard presentations, the reference memory then contains representations of the standard duration that the participant can use (what form the storage might take is discussed later). The next phase is the presentation of comparison stimuli. When a comparison stimulus is presented, it is timed by the clock as before, and is then transferred to working memory. To effect the behavioural response, some sort of comparison and decision process is needed. SET supposes that the content of working memory (essentially the duration of the comparison just presented) is compared with a sample drawn from the reference memory of the standard. The question then is whether the comparison duration is "close enough" to the standard to produce the YES response, or whether it is sufficiently different to warrant a NO response.

Wearden (1992) accounted for temporal generalization in humans using a modified Church and Gibbon (MCG) model derived from an earlier work by Church and Gibbon (1982) on temporal generalization in animals. The details of this and other models are discussed in Chap. 4.

For present purposes, it is sufficient to say that the model has two parameters when fitted to data. One (*c*) is the variability of the underlying representation of the standard, which can come either from variability added from storage in reference memory when the standard duration is learned, or from intrinsic variance from the internal clock itself. Larger values of *c* represent time representations with greater variability from one trial to another. The other parameter is the response threshold, *b*, which reflects how conservative the person is in responding YES. If *b* is large, then even quite large differences between the standard and comparison durations will result in YES responses, whereas if *b* is small, a YES response will occur only when the standard and comparison are close in duration—that is, *b* determines how conservative the decision to respond YES will be.

The MCG model fits data well from most temporal generalization experiments with human participants. It predicts that the peak of YES responding will occur when the comparison duration presented is the same as the standard (mean accuracy), and it also predicts that comparisons longer than the standard by some amount are more likely to be confused with it than those shorter than the standard by the same amount. For example, when the standard is 400 ms, 500 ms produces more YES responses than 300 ms, and 600 ms more than 200 ms (Wearden, 1992). At first sight, this might seem to violate the principle of mean accuracy, as 500 ms "seems closer" to 400 ms than does 300 ms. However, the asymmetry arises from the decision process proposed (as discussed in detail in Chap. 4), not from the underlying representations of time, which are generally accurate. The essential point here is to illustrate how SET's explanation of temporal generalization performance utilizes all three levels of the SET system: clock, memory, and decision.

Models like the MCG model can be used to fit data from different conditions and from different participant groups, and when SET is used, the parameters of the models fitted to data are usually psychologically meaningful. For example, an early use of the MCG model was to compare performance in groups of older people who differed in age or IQ (Wearden, Wearden, & Rabbitt, 1997) as well as to compare the performance of these groups with that of students an average of 50 years younger. If the temporal generalization performance of older individuals differs statistically from that of students, are the differences caused by changes in the variability of time representations with increasing age (*c* in the model) or by changes in decision criteria (*b*), or both? The effects of ageing on timing are discussed in greater detail in Chap. 8, but for present purposes we note that the differences between groups were mainly in *c*, which increased with age, and increased with lower IQ, such that the temporal representations of older people and those with lower IQ appeared more variable than those of younger participants. The important issue here is that modelling consistent with SET can do more than conventional analyses, which merely tell us that groups differ statistically in performance: it can tell us just what the causes of the differences are in psychological terms. The use of SET-consistent models as an account of human timing has been extremely valuable in explaining between-group differences, such as those between groups of children of different ages, as material in Chap. 7 shows.

These examples illustrate the operation of the SET system, showing how clock, memory, and decision processes interact to produce behaviour. It is clear from this material that the prediction of behaviour using the SET system depends heavily on suppositions about how the different parts of the model work, and the choice of decision mechanisms is particularly critical. This has led to the charge that SET is unfalsifiable: if data do not fit one form of the theory, then another can be devised (Staddon & Higa, 1999). Staddon and Higa went so far as to characterize SET as an "erector set" theory, a term intended to be disparaging, claiming that different parts from the "set" could be arbitrarily chosen to fit any data obtained, making it impossible to falsify the model. Although this charge is not without a certain force, it could be equally argued that different timing tasks do indeed involve different mechanisms, thus requiring different components from the "erector set" that SET provides. The problem arises in trying to verify certain principles of the model independently of others.

In an article published around 15 years ago (Wearden, 1999), I tried to address this issue, and suggested what might be called an "isolation" programme, a search for procedures that could manipulate the different proposed components of SET—clock, memory, and decision—separately from the others. This programme has proceeded in a rather disorganized way ever since, but results from it suggest that the main components of SET can be changed individually, thus implying that they have some psychological reality.

Isolating the Pacemaker-Accumulator Clock

The first component of SET that might be usefully isolated is the pacemaker-accumulator clock itself. If a clock of the type proposed by SET exists, at least in functional terms, then one consequence is the type of temporal scaling to be expected in data. If we ignore the time needed to start and stop the clock, a pacemaker-like mechanism would be expected to produce something close to linear time scaling. If the duration timed is doubled, the number of internal clock ticks should also double, and likewise should be halved if the duration is halved. This is true not only of periodic pacemakers, where the time between ticks is constant (as proposed by Treisman, 1963, for example), but is also true on average for pacemakers which pulse at random times but with some average time between ticks. One line of evidence consistent with this notion is the *mean accuracy* so often found in data as to be embodied as a basic principle of SET (Wearden & Lejeune, 2008). However, deciding between linear and non-linear time scales more generally is difficult. Wearden and Jones (2007) provide a detailed discussion of why this is the case, but one method that is useful here is to compare judgements of the duration of *part* of one interval with judgements of the duration of the *whole* of another. Why is this informative? Suppose we contrast two types of pacemakers: one produces a linear time scale, where an interval *n* times as long as another involves *n* times the number of ticks. If the number of pulses reflects subjective time, then with such a pacemaker, when a person is half-way through a time interval in clock time, they should be half-way through in subjective time, and so on, for every subdivision of the interval. In contrast, suppose that when an interval is timed, the pacemaker initially runs quickly and then slows or, conversely, increases in rate as the interval proceeds. Now, when a person is half-way through an interval in clock-measured time, they will not be half-way through subjective time: if the clock slows, the first half of the interval will seem longer than the second half; if the pacemaker accelerates, the reverse is true.

Allan (1979) reviewed a number of experiments involving "ratio setting", where a person had to adjust one interval so that it was some fraction (which could be greater than 1.0) of another. In general, these experiments tended to support linear timing. A more recent experiment, which was essentially an elaborate version of ratio-setting, was conducted by Wearden and Jones (2007). Individuals were given a few examples of a 10-s time interval (not identified as such), which started and ended with clicks, and counting was prevented by a secondary task. After this 10-s interval had been presented, comparison stimuli ranging from 1 to 10 s in 1-s steps were presented in random order, and the participants' task was to decide what percentage each comparison was of the standard. A replication of this procedure, occasioned by a programming error, involved all intervals being 0.5 s shorter than intended (i.e. the standard was 9.5 s long, the comparisons 0.5 s, 1.5 s, etc.). Results are shown in Fig. 3.6. Although the judgements were not perfectly accurate, there was no suggestion of an underlying non-linear time scale, and the small standard errors of the mean indicate that different people in the group were in good agreement as to what the percentage value was in each case. This was despite the participants' initial belief that the task was extremely difficult: in fact, they had few problems with it. Of course, this is exactly the result expected if people possess a mechanism such as a pacemaker-accumulator internal clock which they can access. If start and stop times of the clock are negligible or are small relative to the intervals timed, then doubling or halving a time interval will double or halve the number of pulses accumulated, thus doubling or halving subjective time. For a more complex experiment which tries to distinguish between linear and non-linear time scaling in humans in a different way, but which also concludes that the scale is approximately linear, see Wearden (2002).

Other evidence for a pacemaker-accumulator clock comes from attempts to change the rate of the pacemaker itself. Early work on body temperature and time estimation involved this same idea (Wearden & Penton-Voak, 1995) but in 1990, Treisman, Faulkner, Naish, and Brogan introduced a less technically troublesome method. Treisman et al. (1990)

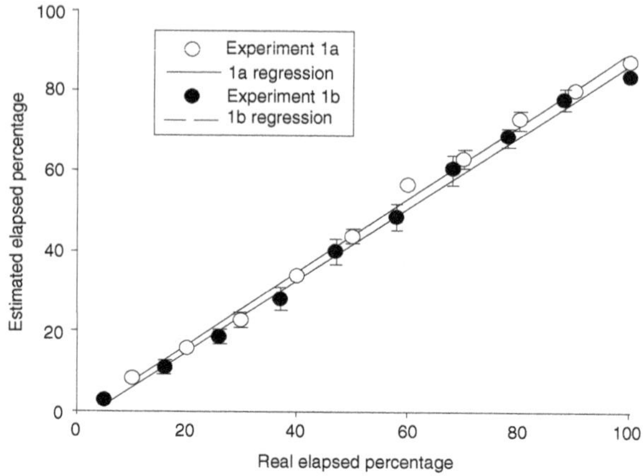

Fig. 3.6 Data from Wearden and Jones (2007). The task was to estimate the elapsed percentage of a standard (in different cases 10 s or 9.5 s) when comparison values varied from 10 to 100 % of the standard. The lines shown are best-fitting regression lines. See text for other details

presented repetitive stimulation in the form of trains of clicks at the same time as visual stimuli whose duration had to be judged, and they found that when the clicks were delivered, participants judged the stimuli as lasting slightly longer than when the repetitive stimulation was absent. The authors proposed that the clicks increased the rate of the pacemaker by increasing arousal, following the original notion of Treisman (1963) that the pacemaker of the internal clock was sensitive to arousal levels. The repetitive stimulation method is particularly useful, as it seems to manipulate only the clock component. If the stimuli to be judged are the same, and the type of judgements made are the same, it then seems that the only source of behavioural differences between conditions with and without repetitive stimulation must be clock operations themselves.

The work of Treisman et al. was followed on by Penton-Voak, Edwards, Percival, and Wearden (1996), who used trains of periodic clicks, usually lasting 5 s, which preceded either auditory or visual stimuli whose duration had to be judged in various ways. They found, for example, that estimates of duration were higher after clicks than without clicks, and that differences in the click/no-click estimates increased with increasing

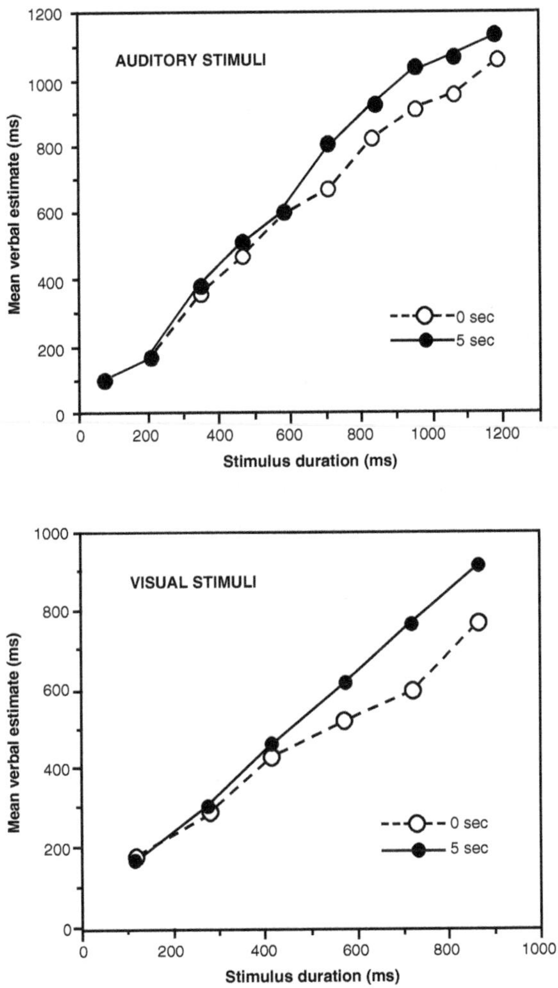

Fig. 3.7 Data from Penton-Voak et al. (1996). *Upper panel*: Verbal estimates of the duration of auditory stimuli (500 Hz tones) preceded by no clicks (0 s) or 5 s of 5-Hz clicks (5 sec). *Lower panel*: Verbal estimates of the duration of visual stimuli (squares of colour on a screen) without clicks or preceded by 5 s of clicks

duration, regardless of whether the stimuli judged were in the auditory or visual modality. Figure 3.7 shows some results from their article.

The mathematics of internal clock theory (Appendix) predict exactly this "slope" effect if the pacemaker has been speeded up. Furthermore, the work of Penton-Voak et al. preceded trials in which participants were required to produce time intervals. Here, click trains made the times produced shorter—again, the result expected if pacemaker speed had increased (Appendix), and in fact the result obtained from Mrs. Hoagland (see Fig. 2.1).

Periodic repetitive stimuli such as trains of clicks or flicker have been used by other researchers, and the effects generally seem robust, with the subjective duration of stimuli appearing to be increased by clicks and flicker. Effects were found in participants ranging from 3- and 5-year-old children (Droit-Volet & Wearden, 2002) to patients in their mid-60s with Parkinson's disease and healthy people of the same age (Wearden et al., 2009). When the judged stimuli themselves consisted of flickering stimuli, subjective durations were once again lengthened (Ortega & Lopez, 2008) compared with static stimuli, and presenting flickering visual displays after stimuli to be judged had the opposite effect, making durations appear shorter (Ono & Kitazawa, 2011). A more recent study by Herbst, Javadi, van der Meer, and Busch (2013) systematically investigated the effects of flicker frequency on changes in subjective duration, in which the time judgement study was accompanied by EEG and measurement of the flicker-fusion threshold, which is the frequency of flicker appearing static to the observer. For the timing task, two stimuli were presented. One stimulus 2 s long flickered at 165.7 Hz, and was perceived as static by all observers. The other flickered at various frequencies (from 3.9 to 165.7 Hz) and had variable durations from 0.5 to 3.5 s. The task was to judge which stimulus lasted longer. Lower flicker frequencies led the participants to judge the flickering stimulus as longer, and the effect of flicker disappeared when the frequency reached the flicker-fusion threshold and was no longer perceived as flickering. This result implies that the *conscious* perception of flicker seemed necessary to lengthen subjective time, and this occurred despite EEG responses still registering the stimulus as flickering even when the flicker rate was above the flicker-fusion threshold.

Although Treisman et al. (1990) proposed that the effects of repetitive stimulation were due to increases in arousal, whether click trains or flicker are arousing in the everyday sense of the word is perhaps questionable. However, the notion of an arousal-sensitive pacemaker has often been influential, and one area where this idea has been frequently applied has been research on emotion, which will be discussed in Chap. 6.

It has long been known that the perception of stimulus duration is determined in part by the nature of the stimulus, and explanations for this type of effect in terms of different pacemaker speeds have recently been proposed. One effect, known since the nineteenth century, is that "sounds are judged longer than lights" (Goldstone & Lhamon, 1974); in other words, auditory stimuli almost always seem to last longer than visual stimuli of the same duration. Another venerable effect is the "filled-duration illusion", the phenomenon by which a time period that is filled (e.g. a continuous tone or visual stimulus) appears to last longer than an unfilled interval (started and ended by clicks or flashes, for example) of the same duration. Figure 3.8 shows data from two experiments where the duration of auditory or visual stimuli (upper panel) or filled and unfilled intervals (lower panel) were estimated.

Both examples clearly show the "classical" effect: "sounds are judged longer than lights" in the upper panel, and filled auditory intervals are judged as markedly longer than unfilled intervals in the lower panel. In both cases, the difference between the auditory/visual or filled/unfilled judgements increases as the judged intervals lengthen, a result closely resembling the "speeding up the clock" effect shown in Fig. 3.7. Wearden, Edwards, Fakhri, and Percival (1998) proposed that the auditory/visual duration judgement difference was due to different pacemaker rates for the auditory and visual stimuli, resulting from a single pacemaker running at different speeds for the different stimuli, or two different pacemakers with different speeds. As shown in the appendix to this chapter, different pacemaker speeds would produce a difference between judgements which would increase as the intervals judged increased, as shown in Fig. 3.8. Wearden, Norton, Martin, and Montford-Bebb (2007) proposed a similar explanation for the filled/unfilled duration phenomenon, once again manifested as a "slope effect", as shown in Fig. 3.8.

This explanation begs the obvious question of why pacemaker speed differs between the two cases, and there is presently no answer. The similarity between the "modality effects" shown in Fig. 3.8 and the click train effects in Fig. 3.7, however, is striking, and differences in pacemaker speed provide a simple explanation that fits the pattern of results obtained.

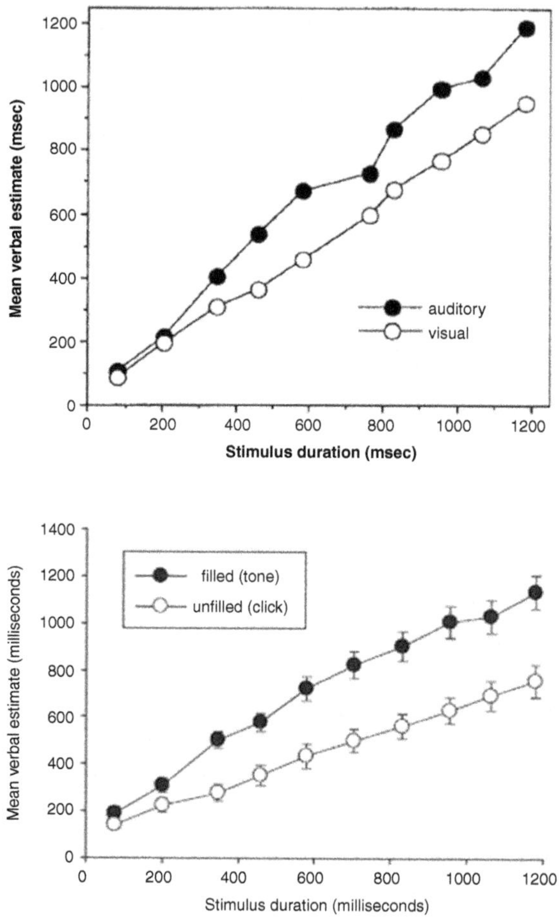

Fig. 3.8 *Upper panel*: Verbal estimation of the duration of auditory (tone) and visual (square on a computer screen) stimuli. Data come from Wearden et al. (1998). *Lower panel*: Verbal estimation of the duration of filled (tones) and unfilled intervals (started and ended by clicks). Data from : Wearden, Norton, Martin, and Montford-Bebb, et al. (2007)

There are many anecdotes in which time experience is reported to change in emergency situations, such as climbers who experience falls or individuals involved in car crashes or other life-threatening situations, and these observations may, of course, be linked to the idea of an arousal-sensitive timing mechanism. Arstila (2012) reviews and discusses many of these anecdotes. The most common report derived from this "emergency time" is that external events seem to slow, in that events seem to last much longer than they are in reality, and may even appear perceptually in a sort of slow motion. This is exactly the effect expected if the speed of an internal clock pacemaker were to be increased dramatically.

There are often other effects as well, such as the experience of heightened sensory clarity. Noyes and Kletti (1976) provide a number of examples. Stetson, Fiesta, and Eagleman (2007) attempted to duplicate this phenomenon by subjecting individuals to more than 2 s of freefall from a tower into a net. However, although the duration of the fall was overestimated, there was no improvement in people's visual resolution when observing a flickering visual stimulus, suggesting perhaps that the proposed effects of real emergencies were not wholly duplicated, possibly because the experimental situation could not create the same feelings as a real emergency.

It seems, then, that exposing people to frightening events in controlled situations does not provide conclusive evidence of changes in subjective time, and there are obvious ethical limitations to studies attempting to increase arousal well above normal levels. In a study in 2008 (Wearden, 2008a), I tried the opposite approach, and sought to *reduce* arousal levels, by the simple expedient of spacing out trials to make the experimental procedures especially tedious, and compared performance early in the experiment with that occurring much later. Obviously, any changes in arousal needed to be verified, and to do this I used a self-rating scale derived from Thayer (1967). The manipulation was effective, in that rated arousal was always significantly lower at the end of the experimental session than at its start, but an additional problem was that the very boring experimental procedure may have produced effects because it reduced participants' attention or motivation to perform the task. This possibility was investigated by contrasting different procedures, two of which will be mentioned here. One was normal temporal generalization. At the start of the procedure, the standard duration was presented, and participants were required to judge whether comparison durations were or were not the standard, with no feedback provided. The logic was that if the standard was initially encoded as *n* ticks of the internal clock, increasingly longer stimuli would then be needed to produce these *n* ticks as arousal diminished with progression of the experiment, and thus longer and longer stimuli would be identified as the standard as the experiment continued. This was the result found. Suppose, however, that two stimuli were presented on the trial, and the task was merely to decide whether they were of the same duration ("episodic" temporal generalization; see Wearden & Bray, 2001). Now, both stimuli on the trial would presumably be timed by the same pacemaker speed, even if this slowly changed over the course of the experiment, and therefore no effect of arousal changes on pacemaker speed would be expected. On the other hand, if attention to the task or motivation to perform it declined as the experiment progressed, performance on the episodic task would then be expected to deteriorate. In fact, no change in the episodic task occurred, although longer durations were identified as the standard at the end of the procedure than at the beginning in normal temporal generalization, suggesting that decreases in arousal "slowed down" the internal clock pacemaker, much as Treisman (1963) suggested.

The research discussed in this section shows that individuals often behave as though they possess a pacemaker-accumulator clock of the type SET proposes. Evidence from ratio-setting experiments suggests that some internal process behaves as though duration is being accumulated, more or less linearly. Furthermore, attempts to speed up or slow down the pacemaker of the clock are naturally presupposed on the idea that some such mechanism actually exists, at least functionally.

Isolating Temporal Memories

Attempts to influence the memory components proposed by the SET system have been much less frequent than those seeking to alter the speed of the pacemaker-accumulator clock, which is perhaps surprising, as one of the memory components, the reference memory, played a critical role in the initial version of SET published by Gibbon, Church, and Meck (1984). SET was initially applied to data from animal experiments, and the FI schedule used with animals, mentioned earlier, can be used to illustrate the importance of reference memory to the early theory. Suppose a rat is working under FI 30-s, so obtains rewards for responding more or less exactly every 30 s. SET originally proposed that the time value associated with reward (30 s in this case) is stored in reference memory, whereas time elapsed in the interval is stored in working memory (thus continually changing as the interval elapses). When a sample from reference memory and the contents of working memory are "sufficiently close", the animal begins to respond in the interval (for a more detailed and formal version of this idea, see Gibbon, Church and Meck 1994).

During training on FI or on other procedures, an animal like a rat or pigeon will receive hundreds or thousands of instances of the time associated with rewarded responses. How are these stored in reference memory? SET originally proposed that each instance of the time of reward is initially timed accurately, but stored inaccurately, after being transformed by a "memory constant". Suppose the interval is 30 s, and this is represented by 300 ticks of the clock (the figure is completely imaginary but helps exposition here). On average, the number of ticks stored in any instance is 300, but can be smaller or larger as the memory constant multiplies the 300 ticks by a random value, which is on average 1.0, but has a Gaussian distribution. So, for example, on one interval, the 30-s period might be stored as 340 ticks, on another 273, and so on. Because of the inverted U-shape of a Gaussian distribution, the majority of the tick values will be close to the average, 300, but some remote values might also occasionally occur. Over a large number of trials, the experimental subject thus builds up a memory distribution of the number of ticks associated with reward, and it is a sample from this distribution which is used on each trial. The animal begins to respond earlier or later in the interval than the average, because the number of ticks retrieved is markedly less or greater than the average. The form of the memory constant proposed, a Gaussian distribution with a mean of 1.0 and some standard deviation, *c*, means that the distribution of the number of ticks associated with reward in memory has a scalar property. In general, the number of ticks would correspond to the time of reward (mean accuracy), but the fact that the Gaussian distribution has a mean of 1.0 and a constant coefficient of variation produces the scalar property in memory as the time of reward changes from one condition to another. This idea is the origin of Gibbon et al.'s (1984) title, "Scalar timing in memory": the scalar property is derived from the properties of the reference memory, not from the operation of the internal clock itself. It is difficult or impossible, however, to distinguish between accurate timing of the time of reward and subsequent "incorrect" storage, and correct storage of times of reward which are associated with different numbers of ticks from one trial to another. Suppose, for example, that the pacemaker of the clock ticks at different rates on different trials, with the rate on each trial sampled from a Gaussian distribution with some constant coefficient of variation. Here, the variation in pacemaker rate is the source of the scalar property, but the memory will contain exactly the same range of values as though the memory "distortion" itself were the source. This is why, in recent expositions of SET, the scalar property is attributed to "memory or timing variability", as the two cannot be distinguished, and I will follow that usage here and later in this book.

Speculation about the contents of temporal reference memory in humans must confront the problem that, unlike the training given to animals, the presentation of standard durations to adult humans usually involves just a few examples, perhaps three or five, so it is hard to imagine that temporal reference memory in humans contains an extensive distribution of hundreds or thousands of items, as proposed for reference memory in animal timing. In that case, what properties does it have? Jones and Wearden (2003) attempted to study the acquisition of "standard" time intervals in a modified temporal generalization task. The new feature of the task was that the "standard" was in force only for a block of seven comparison stimuli ranging from 0.25 to 1.75 times the "standard", which was then replaced by another "standard", and different comparison stimuli, and so on. Participants were informed that the standard changed in each block. In this study, in different conditions, individuals received one, three, or five presentations of the standard at the start of the block. As repetition of items in conventional memory tasks is known to improve memory (Hintzman & Block, 1971), it is perhaps surprising that the temporal generalization gradients obtained did not differ significantly as the number of standards changed over values of 1, 3, and 5. Performance might have been expected to improve (e.g. the temporal generalization gradients would cluster more narrowly around the standard value) with increasing numbers of standard presentations, but this did not happen. This rather counterintuitive result was replicated by Ogden and Jones (2009) using a temporal reproduction task. Jones and

Wearden (2003) used computer simulation to try to explain their findings. The details are too complex for full discussion here, but briefly, they showed that if people stored each standard separately, and then chose one example to serve for the judgement of the comparison stimuli on that block, increasing numbers of standards would then have no effect: only if the standards were averaged together would repetition produce improved performance. This explanation, in fact, returned to the original concept of SET as applied to animals, that examples of the "critical" time are stored separately in a memory distribution.

In a second study, Jones and Wearden (2004) investigated potential effects of remembering two standards in temporal generalization compared with remembering only one. In general, increasing "memory load" in this way might be expected to worsen performance (Underwood, 1969), and indeed, this was the case: temporal generalization gradients were flatter (i.e. discrimination of standard and comparison durations was poorer) when two standards rather than one were remembered. However, this was only the case if the two standards were actually used with their own comparison stimuli: just presenting two standards, but using only one of them for judgements of comparison stimuli, was not effective.

Delgado and Droit-Volet (2007) manipulated temporal reference memory in a different way. In a first experiment, a bisection task was given to children aged 5 and 8 years and adults. Two groups of each age were used. For one group ("fixed"), the standard *Short* duration was 1 s and the standard *Long* was 7 s, and these were presented five times each in a "familiarization" block. For the other group ("variable"), the *Short* and *Long* standards were not, in fact, the same on each of the five "familiarization" trials, and were values chosen from between 0.75 and 1.25 s for the *Short* standard and 5.25 and 8.85 s for the *Long* standard, although the participants were not informed that the standards differed from one trial to another. Following this, test blocks with comparison durations from 1 to 7 s were presented. Psychophysical functions from the variable condition were flatter than for the fixed condition among the children, with significantly higher Weber fractions obtained from the variable group, indicating lower temporal sensitivity.

A second experiment used a temporal generalization method with a standard that was 4 s in a fixed group, but was a value varying from 3 to 5 s for the variable group. These "standards" were presented during a training phase, followed by comparison durations ranging from 1 to 7 s. Temporal generalization gradients were flatter in the variable than the fixed group for the two groups of children, but not for the adults, and modelling using the MCG model (discussed in Chap. 4) suggested that the principal effect of presenting variable "standards" was to increase the value of the timing/memory variability parameter. These two experiments together show that presenting variable "standards" not only decreased timing sensitivity measured empirically in both bisection and temporal generalization—at least when children were used—but also changed the underlying timing/memory variability parameters when modelling was used.

Interference is a common property of memory. Suppose that people learn a series of lists, some containing animal words, others words for furniture. If a retention delay is imposed after learning, then on recall, people are likely to confuse which animal words were in which list, and which words for furniture were in which list: items that are similar interfere with one another. This also seems to be true of memory for duration. Filippopoulos, Hallworth, Lee, and Wearden (2013) interleaved blocks of temporal generalization trials, each beginning with a different standard duration, followed by comparison stimuli which had to be compared with the standard at the start of each block. The experimental blocks were arranged so that medium-duration blocks (average standard duration 400 ms) were interleaved between blocks involving shorter (200 ms) or longer (600 ms) standards. Performance on the averagely constant 400-ms blocks was affected by the values used in the other blocks: individuals tended to decide that shorter comparison stimuli were the standard when the interleaved blocks were 200 ms, and longer when the interleaved blocks were 600 ms. This occurred despite the fact that participants were never required to compare the standards or make comparisons between blocks. Perhaps more surprisingly, the effects occurred when the stimuli in the blocks were of different modalities, either auditory or visual, so auditory standards and comparisons affected judgements of visual stimuli, and vice versa.

The effects in the Filippopoulos study, albeit statistically significant, were small—although it is perhaps remarkable that any effects were obtained at all. The experiment by Ogden, Wearden, and Jones (2008), on the other hand, yielded much more dramatic results. A temporal generalization method was used, and participants received a standard duration followed by comparisons, then a short delay, then another standard (either shorter or longer than the first), with its own comparison stimuli. Next, a delay was imposed, and then the comparison stimuli for the first standard (but not the standard itself) were presented again. The second standard/comparison block after the delay had a dramatic effect on performance. It seemed as though memory of the initial standard had been completely erased, although participants seemed to remember the *relation* between the stimuli in the two blocks, that is, whether the second block involved longer or shorter stimuli. This produced markedly distorted temporal generalization gradients, although a version of the MCG model with "memory distortion" (discussed in Chap. 4) fitted gradients well despite their highly abnormal forms at first sight. Figure 3.9 shows the results. Unconnected points show the temporal generalization gradients, which in some cases were skewed towards the longest or shortest comparison

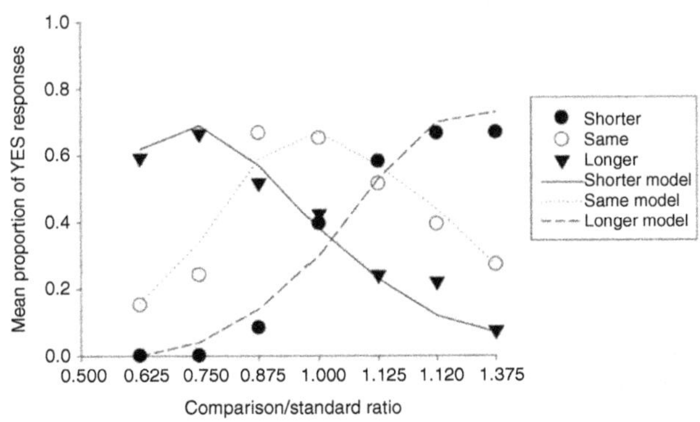

Fig. 3.9 Data from Ogden et al. (2008). Temporal generalization gradients are shown from conditions where an interfering duration set was shorter, longer, or of the same duration as the standard duration/comparison set tested. Unconnected points show data, and the lines show the best fit of a variant of the MCG model with memory distortion

stimuli presented, and the fit of the MCG model with "memory distortion" is shown as well. These two experiments seem to show that interference between durations stored in memory can occur in the same way as interference in memory between other sorts of items such as words.

The experiments just discussed involved *reference memory* for time, as the interference occurred between standards that were valid for a block of trials, whereas a few studies have investigated the properties of *working memory* for time, and these have uncovered an unusual phenomenon, *subjective shortening*. This was popularized initially as the result of an experiment by Spetch and Wilkie (1983) with pigeons, which were tested in an operant chamber with three keys. This experimental procedure, in simplified form, was as follows. With the two side keys dark, the centre key lit up for one of two durations (2 or 10 s). When the centre key went dark again, the side keys were illuminated, and the pigeon was rewarded with food for a peck on one of them (e.g. a red key) if the duration had been 2 s, and on the other (e.g. green) key if the duration had been 10 s. When the animals learned to discriminate the centre key illumination durations well, a delay was introduced between the centre key darkening and the side keys lighting up. If we call pecking the key associated with the longer duration the "choose long" response, and the other the "choose short" response, Spetch and Wilkie found that pigeons showed an increasing tendency to "choose short" as the interval between the end of the presentation of the sample duration on the centre key and the opportunity to respond on the side keys increased. They attributed this "choose short effect" to a phenomenon they called *subjective shortening*. The longer the duration of a stimulus was retained in working memory, the shorter it seemed to be. An analogy might be the retention of liquid in a sieve: the longer the liquid is retained, the less there is at the end of the retention interval.

A method like that used by Spetch and Wilkie (1983) cannot be employed with humans, who would simply verbally label the sample durations as "short" or "long", and retain this label. Wearden and Ferrara (1993), however, developed a technique for investigating potential subjective shortening effects which was usable with humans. Individuals received two stimuli on each trial: the first was called the sample (*s*), the second the comparison (*c*). These were separated by an *s–c* delay, timed from the offset of *s* to the onset of *c*. When *c* had been presented, participants

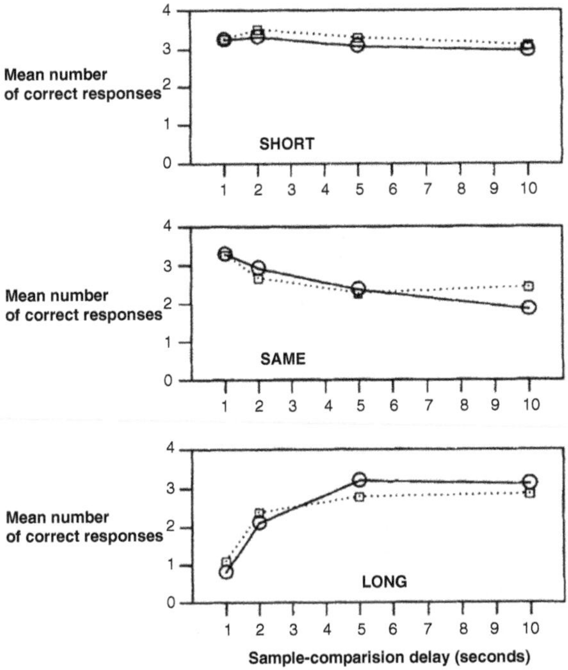

Fig. 3.10 Data from Wearden and Ferrara (1993). The mean number of correct responses on the memory for duration task is plotted against sample-comparison delay. SHORT trials: comparison is shorter than the sample. SAME trials: comparison duration is the same as the sample. LONG trials: comparison duration is longer than the sample. See text for other details

were asked whether the second stimulus (*c*) was longer, shorter, or of the same duration as the first (*s*). The logic was that when the comparison is made after *c*, the memory of *c* is fresh, whereas the memory of *s* may be subjectively shortened as the *s*–*c* delay increases. There were three types of trials: *equal/same* trials where $s = c$, *short* trials where $s > c$, and *long* trials where $s < c$. The trials are identified here by the correct response. Wearden and Ferrara found that the effect of increasing the *s*–*c* delay was not the same on all trials. The results are shown in Fig. 3.10. With *short* trials, increasing the delay made little difference to performance, whereas on *equal/same* trials, increasing the delay resulted in increasingly inaccurate performance. However, the most remarkable result was that increasing the *s*–*c* delay led to better performance on *long* trials (where *c* really was longer than *s*). This result is easy to understand if *s* is increasingly subjectively shortened as the *s–c* delay lengthens: *s* becomes subjectively shorter and shorter, and therefore easier to discriminate correctly from *c*, as the *s–c* delay increases.

The subjective shortening hypothesis also predicts the pattern of errors. The easiest case to understand is that of errors on *equal/same* trials, where $s = c$. Suppose that the participant makes a mistake on one of these trials. Subjective shortening predicts that there should be an increasing proportion of "long errors" (saying that *c* is longer than *s*) and a decreasing proportion of "short errors" as the *s–c* delay increases, and this was observed as well. The pattern of results from Wearden and Ferrara's procedure is complex, and the interested reader is referred to the original article and those discussed below. For present purposes, however, we should note the two main "signatures" of subjective shortening with this method: the increasing accuracy of responding with *s–c* delay on *long* trials, and the pattern of errors on *equal/same* trials.

Wearden and Ferrara found evidence for subjective shortening in humans' memory for stimulus duration, yet a question arises is to whether subjective shortening is restricted to duration judgements, or whether it is also shown in remembering other aspects of stimuli. What is needed as a comparison for duration is some other stimulus aspect that cannot be immediately given a verbal label, and which can be varied quantitatively. Wearden, Parry, and Stamp (2002) contrasted *duration* with *physical length* (hereafter in this section simply "length"). The stimuli were line-like thin bars of colour. When two were presented successively, questions could be asked about their respective duration (as in Wearden & Ferrara, 1993) or about their length (i.e. was the second line longer, shorter, or of equal length to the first). In fact, both length and duration were varied from one trial to another, but the participants had to make decisions about only one of them. If the stimuli were presented in an arrangement like that of Wearden and Ferrara, the effect of varying *s–c* delay on duration and length judgements could be observed. Wearden et al.'s (2002) study comprised three experiments, and the results were highly complex overall, but in summary, the two signatures of subjective shortening mentioned above were found only when *duration* and not (physical) *length* was the basis of the decision. This was true regardless of whether the discrimination of duration or length was the easier of the two, and was also found when the stimuli presented in the duration and length discriminations on average were physically identical. It seems, therefore, that not all quantitative stimulus dimensions that cannot be verbally labelled show subjective shortening, and that perhaps the effect is caused by the manner in which duration is represented and stored.

Wearden, Goodson, and Foran (2007) followed on these studies by varying the type of stimulus and simplifying the procedure. In different conditions, stimulus pairs involved filled or "empty" auditory or visual intervals, started and ended by clicks or flashes. The procedure involved *short*, *long*, and *equal/same* trials, but the last type were most common. In addition, the only responses permitted were SHORT and LONG, and thus all responses on *equal/same* trials were errors. On these trials, there was an increasing tendency towards a LONG response as the *s–c* delay lengthened, irrespective of the stimulus type.

Although Wearden et al. (2007) found subjective shortening across all stimulus types, and Wearden and Ferrara's initial (1993) demonstration used tones, a more recent paper by Takahashi and Watanabe (2012), which used a technique just slightly different from that of Wearden and Ferrara, was not able to obtain subjective shortening effects when the sample and comparison stimuli were auditory, although the effect was present with visual stimuli. In other experiments, the sample and comparison stimuli were of different modalities, either visual or auditory. When the comparison stimulus was visual, the subjective shortening effect occurred, but not when it was auditory. Furthermore, there was no effect of delay period when participants did not know in advance what the modality of the comparison would be.

Supposing that subjective shortening is a reliable phenomenon, which my and my colleagues' studies suggest that it is, can it be detected in other types of experiments? One might suppose that subjective shortening effects would "contaminate" procedures like temporal generalization and bisection, although participants are normally judging the comparison stimuli immediately after their offset, or even before (Klapproth & Wearden, 2011), and so effects would perhaps be undetectably small. Wearden and Bray (2001), however, did report what looked like subjective shortening effects in a variant of bisection that they called "episodic bisection". To simplify somewhat, on each trial, individuals received a *short* standard, a *long* standard, and then a comparison which they had to classify in terms of its similarity to the standards. The standard and comparison stimuli varied from one trial to another, and the order of the *short* and *long* standards was randomized from trial to trial. Wearden and Bray (2001, pp. 302–303) noted that when the *short* standard came first, a longer period of time elapsed between the offset of the short stimulus and the end of the comparison than when the *long* standard came first. Over all the trials, this produced the possibility that memory of the *short* stimulus was affected to a greater degree by subjective shortening than memory of the *long* stimulus, thus making the comparison stimulus seem more like the *long* standard, and shifting the bisection point to the left compared with "normal" bisection, and this exact result was found when the standards on average were 300 and 1200 ms long.

Decision Processes

We saw in the preceding section how the memory components proposed by SET can be manipulated. Decision processes can also be altered experimentally. In some ways this is commonplace in the use of SET, as different decision processes are used to model performance on different timing tasks, as shown in Chap. 4, but perhaps a more convincing demonstration of the manipulation of decision processes, independently of other parts of the proposed timing system, comes from an experiment by Wearden and Grindrod (2003). The task used was temporal generalization. Here, there are two sorts of correct responses: a comparison duration which is the same as the standard can receive a YES response, and a comparison stimulus different from the standard can receive a NO response. Wearden and Grindrod awarded points for both types of correct responses, but in different conditions rewarded either correct YES or correct NO responses differentially. Not surprisingly, more points for correct YES responses elicited a greater frequency of YES responses, even when stimuli in the conditions compared were identical. However, modelling showed that this manipulation affected only the threshold for responding: memory/timing variability (*c* in the MCG model; see Chap. 4) was unaffected. Figure 3.11

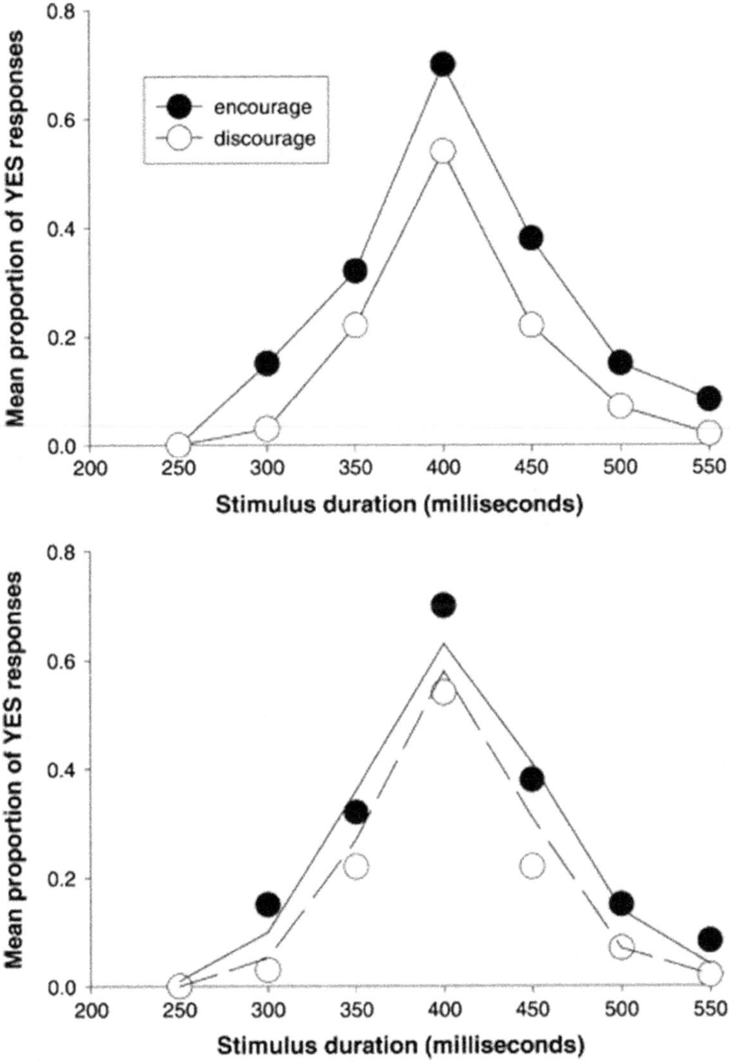

Fig. 3.11 Data from Wearden and Grindrod (2003). *Upper panel*: Temporal generalization gradients from a task with a 400-ms standard and comparison durations ranging from 250 to 550 ms. Encourage condition: more points given for correct YES responses than correct NO responses. Discourage condition: more points given for correct NO responses than for correct YES responses. *Lower panel*: Fits of MCG model to data

shows some of the results. This manipulation might be expected to affect only decision processes (as the standard and comparison stimuli were the same in the conditions compared, so clock and memory processes would presumably be the same), and indeed it affected only the decision parameter in the MCG model, demonstrating the independence of decision processes from the other parts of the SET system.

Another manipulation which altered decision processes while keeping other parts of the SET system constant was discovered by Ferrara, Lejeune, and Wearden (1997). The task used was temporal generalization, with a 600-ms standard, and in two conditions, comparison stimuli were spaced in 150-ms or 75-ms steps around this standard value. These two sets of comparison stimuli had certain values that were physically the same (450, 600, and 750 ms), as was the standard duration, but fewer YES responses occurred at these durations in the difficult (75 ms) set than in the easier one. When the MCG model was used to fit data, the principal change was in the decision threshold parameter, implying that individuals were more conservative with YES responses in the 75-ms set. Paul, Wearden, Bannier, Gontier, Le Dantec, and Rebai (2011) replicated this result almost exactly, and once again the explanation of the behavioural effect which best fitted the data was a change in threshold in the more difficult condition.

The above material shows that the different components proposed by the SET model (clock, memory, and decision processes) appear to be individually manipulable by different experimental operations or different experimental conditions. As I note elsewhere (Wearden, 2013b), this does not prove that SET is correct, but it does seem to suggest that the different components of the timing system are independent in some cases, such that the mechanism generating initial "raw" time representations seems to be separate from the mechanism storing these representations, irrespective of whether these are standards or comparisons, and finally, that some decision mechanism—again, separate from the initial timing and storage mechanism—operates on them to produce behavioural responses.

Summary

SET is a multi-process model of timing, involving a pacemaker-accumulator clock, working and reference memory stores, and decision processes, and resembles the earlier model of Treisman (1963). Different decision processes are used to model performance on different timing tasks, and quantitative modelling gives rise to parameter values which are generally psychologically meaningful. One of these is the measure of the variability of time representations, which can derive from internal clock or memory processes. The other is often some sort of threshold for making behavioural responses, reflecting decision processes. Modelling using SET can account for performance differences between groups (people of different ages, for example) in terms of underlying psychological processes. The different components of the SET model (clock, memory, decision) can apparently be manipulated separately. Some studies use trains of clicks or flashes to "speed up" the pacemaker of the proposed internal clock, and the concept of different internal clock pacemaker speeds has been useful in attempts to understand why time judgements are affected by stimulus type. In other studies, the contents of reference memory and working memory have been manipulated. Finally, some experiments show that decision processes can be changed by rewarding responses or by varying task difficulty. Taken together, these results suggest that SET's division of timing processes into clock, memory, and decision components has some psychological reality.

Simple Mathematics of Pacemaker-Accumulator Clocks

In this section, I provide some simple mathematics of a pacemaker-accumulator clock like that proposed by SET, and in particular, I want to illustrate what might be expected if experimental manipulations are claimed to change pacemaker speed or alter switch processes. The sole focus of interest here is the number of clock ticks accumulated in various conditions, which is considered to represent the "raw material" for time judgements.

Suppose that we have a pacemaker that ticks on average at r ticks per second. When an event to be timed begins, the switch closes with some non-zero latency l_c, so pulses begin to flow, and when the event to be timed ends, the switch opens again, with some non-zero latency l_o. This means that the effective duration of the event or stimulus lasting t seconds is

$$t + l_c - l_o \tag{3.1}$$

such that the total number of ticks accumulated, P, is just

$$P = r\left(t + l_c - l_o\right) \tag{3.2}$$

To simplify, suppose that $l_c - l_o = d$, in that d represents the difference between the closing and opening latencies of the switch, and d can be negative, of course, if the opening latency is greater than the closing latency. This means that

$$P = r\left(t + d\right) \tag{3.3}$$

As t is varied between conditions, the number of accumulated ticks, P, is a linear function with slope r and intercept rd.

Suppose that a claim is made that the pacemaker rate has been changed by some manipulation. How should P change? Suppose that we have two pacemaker rates, r_1 and r_2, and we can arbitrarily define their relation as $r_1 > r_2$. Now, we use Eq. (3.3) to derive P_1 and P_2, the number of ticks accumulating with pacemaker rates r_1 and r_2.

From Eq. (3.3), $P_1 = r_1(t+d)$ and $P_2 = r_2(t+\mathrm{d})$, but we are probably interested in the *difference* between the number of ticks accumulated in the two cases, $P_1 - P_2$. This is

$$P_1 - P_2 = \left(t + d\right)\left(r_1 - r_2\right) \tag{3.4}$$

Equation (3.4) shows that the difference between the number of ticks accumulated is dependent on the difference between the rates of the pacemakers ($r_1 - r_2$), but also on the quantity ($t + d$). If d is constant as the time, t, to be judged changes, then the quantity ($t + d$) is dependent on t. As t becomes larger, the difference $P_1 - P_2$ also increases, and so the effects of the changing pacemaker speed should be greater at longer times than those that are shorter.

For readers who do not like algebraic solutions, a numerical example may help. The numbers used here are completely imaginary, and employed simply for purposes of illustration. Suppose that it takes 20 ms to close the switch to start timing, but 40 ms to open it again to stop, and so $d = 40 - 20 = 20$ ms. This adds 20 ms to the effective duration of the stimuli. To illustrate, suppose we have event durations of 200, 400, 600, 800, and 1000 ms, and two pacemaker rates, 100 and 150 ticks per second. The number of ticks accumulated is shown in Fig. 3.9. It is clear that the pacemaker rate difference manifests itself in a difference in slope in the conditions which are compared. The difference in the number of ticks accumulated between the two conditions increases as the duration timed lengthens: for example, the difference is 11 ticks at 200 ms but 51 at 1000 ms.

If we have a range of t values, then the manner in which the difference in pacemaker speed is manifested in behaviour depends on the task required. If a verbal estimation procedure is used, and we assume that the number of ticks accumulated is directly scaled into estimates, then the pacemaker speed difference should manifest itself as a difference in *slope* when the mean estimates from different t values with putatively different pacemaker rates are plotted against duration, with the higher pacemaker speed giving longer estimates, like the results shown in Fig. 3.12. Some experimental data support this contention (e.g. Penton-Voak et al., 1996; see Fig. 3.7). In contrast, if a method like bisection is used, then the difference between bisection points between the conditions reflects pacemaker rate, so this difference should increase as the stimuli in the bisection tasks increases in length. For example, if we have a 200/800-ms and a 400/1600-ms bisection pair, the difference between the bisection points derived from the conditions with different pacemaker speeds should then be greater in the 400/1600-ms than the 200/800-ms case. Again, there are data suggesting that this is true (Droit-Volet & Wearden, 2002). For another example using a production method, see Burle and Casini (2001; data in their Fig. 3.7, p. 202).

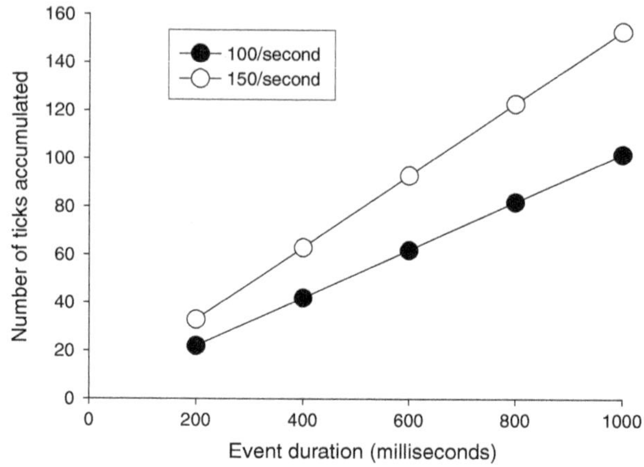

Fig. 3.12 Number of clock ticks accumulated for two different pacemaker rates (100 and 150 ticks per second) for event durations from 200 to 1000 ms. See text for other details

In contrast to changes in pacemaker speed, suppose we have some manipulation which changes the latencies to start and stop timing, *d* in our equations above. This change in start/stop latency is not dependent on the interval timed, and so should have a constant effect regardless of the duration to be judged. In terms of plots such as that in Fig. 3.9, this would manifest itself in a change in *intercept* rather than slope, or a constant difference in bisection point, irrespective of the range of intervals used.

This is easy to demonstrate numerically. Suppose that *d*, the difference between latencies to stop and start timing, changes from 20 ms, as in our example above, to 50 ms. If the pacemaker rate is 100 ticks per second, the effective duration of the stimuli timed then increases by 50 ms, or 5 ticks. Therefore, 25 rather than 22 ticks represents 200 ms, and 105 rather than 102 ticks represents 1000 ms. When the number of ticks accumulated is plotted against duration, the lines from the two conditions compared are parallel. Some studies report changes in condition, usually involving "attentional" manipulations, which show such constant effects. See, for example, Droit-Volet (2003).

In general, then, changing the pacemaker speed produces multiplicative effects on time judgements, such that effects are larger at longer times than shorter times, whereas changing the balance of start/stop latencies by changing switch processes produces constant effects, independent of the interval timed. Given that the idea of an arousal-sensitive pacemaker has been a popular one, multiplicative effects are sometimes referred to as being based on "arousal", whereas constant effects are based on "attention" (see Burle & Casini, 2001, for discussion). However, some theories of attention in timing permit multiplicative effects, and so are difficult or impossible to distinguish from arousal effects, as discussed in Chap. 5.

The pacemaker of the internal clock is usually assumed to be a Poisson timer, in that ticks are emitted at random with a constant average rate. With this type of timer, the number of ticks grows linearly with real time, but it is the variance which increases linearly with real time rather than the standard deviation. This means that a relative measure of time such as the coefficient of variation (standard deviation/mean) will *decrease* as the time intervals judged lengthen, violating scalar timing. However, an underlying Poisson timer can be reconciled with scalar timing in various ways. Perhaps the easiest is to assume that the pacemaker is a Poisson timer, but that the pacemaker rate varies from trial to trial, and on each trial takes a value sampled from a Gaussian distribution, with a mean and some coefficient of variation. Such a process will generate average pacemaker output with scalar properties. Other ways of reconciling scalar timing with an underlying Poisson process are discussed in Chap. 9.

Poisson timers also have the property in which the standard deviation of the number of ticks produced grows as the square root of the pacemaker rate. This means that faster pacemakers produce relatively lower variability: the variability is a smaller fraction of the mean as the pacemaker rate increases. The implications of this property for measures of behaviour are discussed in Wearden and Jones (2013).

4

Theoretical Models of Temporal Generalization and Bisection in Humans

As noted in the previous chapter, one advantage of SET as an account of human timing is that it can give rise to quantitative theoretical models of performance on timing tasks. Furthermore, these models generally have psychologically meaningful parameters, and therefore can account for differences in the behaviour among different groups or the same groups in different conditions in terms of differences in underlying psychological processes. This chapter introduces some of the commonest theoretical models used over the last 20 years or so, discusses the ways in which they were developed, and outlines what seem to me to be their strengths and weaknesses. Temporal generalization and bisection have been chosen here because (a) well-developed theoretical models exist for both of them, and (b) quantitative models have been fitted to data in a large number of experiments, as later chapters will show.

J. Wearden, *The Psychology of Time Perception*,
DOI 10.1057/978-1-137-40883-9_4

Temporal Generalization Models

Temporal Generalization with Standards (Wearden, 1992)

The original temporal generalization task was developed in an experiment by Church and Gibbon (1982) using rats as subjects. Rats were rewarded with food for a lever-press response after a stimulus of a standard duration (e.g. 4 s), but responses were not reinforced if the stimulus had a longer or shorter value. When the probability of responding was plotted against stimulus duration, the function resembled a Gaussian distribution centred on the standard value, with response probability falling off symmetrically as the stimulus duration deviated from the standard. The basic equation developed by Church and Gibbon (1982) to account for performance proposed that the rats responded if

$$abs\left(s^{*}-t\right)/s^{*}<b^{*} \quad (4.1)$$

Here, s^* is a sample drawn from a Gaussian distribution with a mean equal to the standard s and some coefficient of variation, and b^* is a sample drawn from a Gaussian distribution with mean b and some coefficient of variation. Here, b is a threshold which determines how "close" s^* and t must be to generate a response, in which t is the stimulus duration just presented, assumed to be timed without error, and *abs* indicates absolute value (i.e. difference regardless of sign).

Equation (4.1) generates symmetrical generalization gradients, such that some stimulus duration x s longer than the standard s generates the same proportion of responses as a stimulus x s shorter. The model that Church and Gibbon (1982) fitted to the data also had an "inattention" parameter, needed to reflect the fact that rats responded to stimulus durations that were remote from the standard.

Wearden (1992) and later researchers typically found that temporal generalization gradients from humans were asymmetrical, and right-skewed, such that a greater number of YES responses (judgements that a comparison duration, t, was the same as the standard, s) occurred for stimuli longer than the standard by some amount than for those shorter by the same amount: for example, if $s = 400$ ms, more YES responses occurred after a 500-ms stimulus than a 300-ms one. For an example with longer stimulus durations, see Fig. 3.4.

To account for gradient asymmetry, I kept the basic structure of Church and Gibbon's model, but modified the equation slightly, to produce what I called the MCG (modified Church and Gibbon) model. The basic rule was to respond YES if

$$abs\left(s^{*} - t\right) / t < b^{*} \tag{4.2}$$

Here, all the terms have the same meaning as in Church and Gibbon's original proposal. The only change between their original model and mine was that the difference between s^* and t was divided (or "normalized," as Church and Gibbon would say) by t rather than s^*. In other words, the difference was expressed as a fraction of the just-presented duration t rather than s^*. This modification produced asymmetrical generalization gradients skewed to the right, and generally was well-fitted to data from adult humans.

To illustrate why the asymmetry occurs, let us ignore the fact that some values are variable from trial to trial and just use means. Suppose that the standard duration is 400 ms and that two comparison stimuli are 100 ms above and below it, 500 and 300 ms. Using Eq. (4.2) above, the resulting value is *abs*(400–500)/500, which equals 0.2 for the 500-ms comparison, and *abs*(400–300)/300, which equals 0.33 for the 300-ms standard. The smaller of the two values (0.2, coming from the 500-ms comparison) will be judged as "more similar" to 400 ms according to the equation above, even though the absolute differences between the standard and comparison stimuli are both 100 ms.

In several articles, the MCG model has been embodied in a computer programme and fitted to data, in the first instance in Wearden (1992). Fits of the model yielded three parameter values. One is c, the coefficient of variation of the representation of s, which embodies memory/timing variance; a second is b, the mean threshold for the YES response; and the third is a coefficient of variation of this threshold. In more recent fits of the MCG model to data, it has been assumed that the coefficient of variation of *b* is 0.5, meaning that the standard deviation of *b* is half the mean, so two free parameters, *c* and *b*, are derived.

In the 1992 temporal generalization article, I also explored two variants of the MCG model. One was exactly the same as the original, except that the threshold had no trial-by-trial variability ("fixed-threshold model"). This model fitted the data slightly more poorly than the basic MCG model, although the fit was still good by most standards in psychology. Another variant explored involved adding trial-by-trial variability to *t* as well as *s*, in which *t* and *s* were both represented as Gaussian distributions with means *t* and *s*, and some coefficient of variation, which was forced to be the same. This variant fitted data about as well as the simple MCG model.

Comparison between the MCG model and the variant with variability added to *t* is informative regarding the operation of this type of model in general. Table 1 of the 1992 article shows parameter values from fits of different models. The *c* value needed for the basic MCG model was between 0.16 and 0.20 for different conditions, whereas when variability was added to *t*, the values declined to between 0.10 and 0.12, but this is only to be expected, as now there are two sources of scalar variance in the timing system rather than one. It is quite common in modelling with SET to "absorb" all the scalar variance in one component of the model—for example, in the reference memory component as in the MCG model—and leave the comparison stimuli variance free. While this may be implausible psychologically, it has little effect on the goodness of fit of the models, although parameter values will be changed as sources of variance are added and subtracted.

The MCG model was developed by Droit-Volet, Clément, and Wearden (2001) to enable it to fit generalization gradients from young children. These had certain characteristics differentiating them from those obtained with adult participants. For one thing, they were less clearly rightward-skewed, and in the case of very young children were leftward-skewed. In addition, very young children produced generalization gradients with a high proportion of YES responses at stimulus durations remote from the standard. Droit-Volet et al. (2001) varied the MCG model in two ways. Firstly, they introduced a "distortion" parameter, following an earlier suggestion by McCormack, Brown, Maylor, Darby, and Green (1999).

This was a parameter, *k*, which multiplied the standard, *s*, so that this became *ks*. If *k* was less than 1.0, then the standard was remembered as shorter than it really was; if *k* was greater than 1.0, the standard was remembered as longer. In addition, a "random responding" parameter was added, something rather similar to the "inattention" parameter used by Church and Gibbon (1982) to model generalization gradients from rats, which proposed that in some proportion, *p*, of trials, the YES or NO response was generated with equal probability without regard to the stimulus duration presented. Results from this type of model are discussed in Chap. 7, but Fig. 4.1 (taken from Droit-Volet et al., 2001) shows the effects on generalization gradients of varying the various parameters of the model, with simulations generated by the MCG model and its variants. The situation simulated is a temporal generalization task with a 4-s standard and comparison durations ranging from 1 to 7 s.

Panel a shows the effect of varying *c* over values of 0.3 and 0.5, with *b* constant at 0.2, $k=1$ (i.e. no distortion) and $p=0$ (no random responding). The effect of flattening of the gradients with increasing *c* is

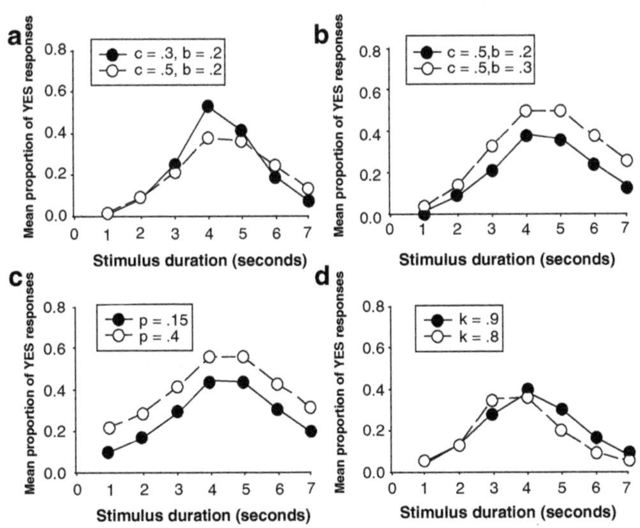

Fig. 4.1 Effects of varying the parameters *c*, *b*, *p*, and *k* of the variant form of the MCG model. See text for details

the principal effect here. Panel b shows effects of varying the threshold *b* over values of 0.2 and 0.3, with 0.3 being a “less conservative” decision to respond YES. Once again, $k = 1$ and $p = 0$, and *c* in this example is kept constant at 0.5. The effect of increasing the proportion of YES responses to all stimulus values is the obvious effect of increasing *b*, although the increase is more marked for longer stimulus durations than for shorter durations. Panel c shows the effect of increasing the probability of a “random” response (*p*) over values of 0.15 and 0.4. Like the effect of increasing *b* (panel b), increasing *p* increases the proportion of YES responses at all stimulus durations, but the effect is not identical to that in panel b, as the effect is the same at all stimulus durations, even short durations remote from the standard, 4 s. Finally, panel d shows the effect of varying the “distortion” parameter, *k*, over values of 0.9 and 0.8. Decreasing *k* systematically shifted the temporal generalization gradients to the left, so the maximum proportion of YES responses could occur at non-standard comparison durations. Increasing k above 1.0 would have the opposite effect, that of increasing the skew of the gradient to the right.

Episodic Temporal Generalization (Wearden, 2004)

Wearden and Bray (2001) introduced an “episodic” variant of temporal generalization which is best illustrated by the procedure of their Experiment 3. On each trial, a “standard” in the form of a tone was randomly drawn from one of three distributions, representing the duration ranges 350–500, 450–750, or 600–1000 ms. When the “standard” duration was selected, a “comparison” stimulus was created which was some multiple of the “standard,” with the multiples being 0.25, 0.5, 0.75, 1, 1.25, 1.5, and 1.75. The order of the “standard” and “comparison” was then randomized, and the two stimuli were presented. The participant’s task was to judge whether they were of the same duration, making a YES/NO response. The terms “standard” and “comparison” are in inverted commas here because, of course, there is no standard for the task in the sense used in normal temporal generalization. Durations were never repeated except by chance, but by plotting the proportion of YES responses against the multiplier value (0.25–1.75), temporal generalization gradients can be produced for the different duration ranges, which can also then be plotted on the same scale to test superimposition.

The MCG model and its variants cannot be used to model behaviour on this task, as there is no essential difference between "standards" and "comparisons." In Wearden (2004), I proposed a model of the following sort. Suppose that t_1 and t_2 are the two stimulus durations presented on a trial. I first added trial-by-trial variability in the usual way, by transforming them into Gaussian distributions with means t_1 and t_2, and some coefficient of variation. Samples from these distributions $t_1{}^*$ and $t_2{}^*$ are used on the trial, and m is their mean. The rule I proposed was to respond YES when

$$\text{abs}\left(t_1^* - t_2^*\right) / m < b^* \tag{4.3}$$

where b^* is a threshold variable from trial to trial around a mean value b, and essentially is used to decide whether t_1 and t_2 are "close enough." The model fitted data well, and bears some similarity to the models used for normal temporal generalization, as well as to the Wearden and Ferrara (1995) bisection model. Issues of consistency between models are discussed later.

Bisection Models

Note: In the discussion of procedures and responses below, italicized letters (e.g. *S* and *L*) indicate stimulus values, whereas non-italicized letters (S and L) indicate responses.

Another common timing task associated with SET is *temporal bisection*. Here, individuals are initially presented with two standard durations, one *Short (S)* (e.g. 200 ms) and one *Long (L)* (e.g. 800 ms). If these are easily distinguishable, as in the example values given, adults need little or no training.

Following presentation of the standards, comparison durations are presented, which usually include the standards and the values between them; for example, durations from 200 to 800 ms in 100-ms steps would be presented in random order. After each comparison has been presented, the participant must judge whether it is more similar to the *Short* or *Long* standard, as indicated by SHORT (S) and LONG (L) responses. Data are typically presented in terms of a *psychophysical function* in the form of the proportion of L responses plotted against comparison duration. Examples from Wearden (1991b) are shown in Fig. 4.2.

The ratio of the *Long* and *Short* standards (the *L/S* ratio) can be considered a measure of the difficulty in discriminating between *S* and *L* and of deciding whether any particular comparison stimulus is closer to *S* or to *L*, as all the comparisons are close in duration, as are *S* and *L* themselves. When the *L/S* ratio is small, participants may receive initial training to ensure that the standards are reliably distinguished, and this training may be extensive in cases where children are used.

With adult participants and a 200/800-ms pair, the psychophysical function rose from near zero L responses at 200 ms to near 100 % at 800 ms (upper panel of Fig. 4.1). The psychophysical function can be analysed to yield a number of measures, but two are of particular interest. One is the *bisection point* (BP), which is the comparison value eliciting 50 % S and L responses. The other is the *Weber ratio* (WR), which is a measure of how steeply the curve rises between comparisons which are the same as *S* and *L*. In some recent experiments, other measures of the slope of the curve are used, but the focus of interest is still on the steepness of the curve gradient in different conditions. The gradient of the curve, the WR or some other measure, reflects temporal sensitivity. If participants are highly sensitive to duration, the curve should rise steeply; if their ability to discriminate durations is poorer, the curve should be flatter. Thus, curve steepness is a measure rather like *c* in the MCG model or the Weber fraction derived from older studies of time discrimination, and reflects underlying timing variability.

What does the BP reflect? It seems reasonable at first sight to propose that the BP is the psychological "mid-point" of the duration range between the *Short* and *Long* standards, just as some point on a trip will be equidistant from the starting point and the destination. In experiments with animals, the BP was almost always at the geometric mean of *S* and *L* (the square root of the product, 4 s for an *S/L* pair of 2 and 8 s, for example, or 400 ms for a 200/800-ms pair). Early experiments with humans produced mixed results, with some experiments (Wearden, 1991b) finding the BP near the arithmetic mean of *S* and *L* ($[S+L]/2$: 5 s or 500 ms in the

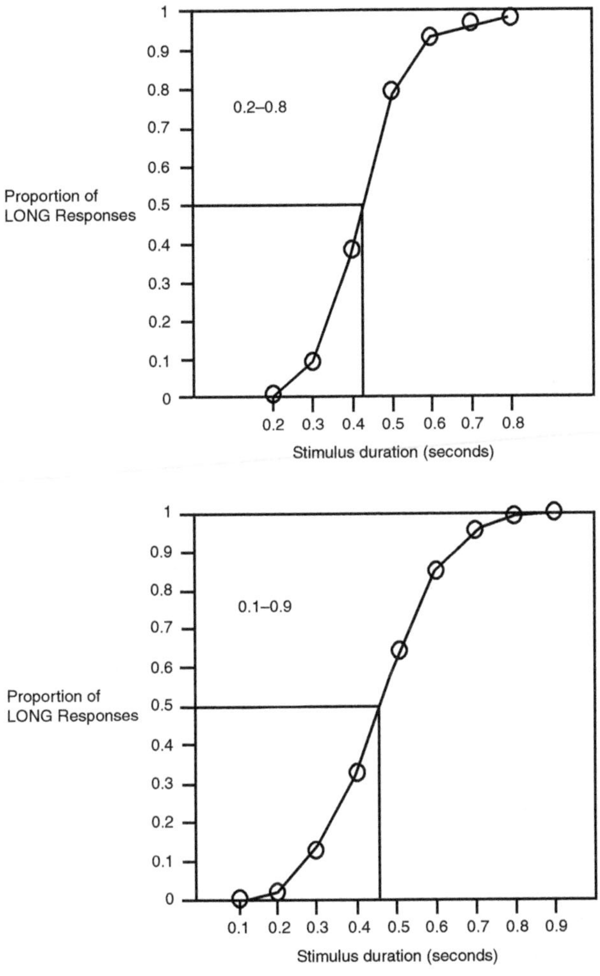

Fig. 4.2 Psychophysical functions from Wearden (1991b). The proportion of LONG responses is plotted against comparison stimulus duration. *Upper panel* shows bisection with 0.2-s and 0.8-s *S* and *L* standards, respectively, and *lower panel* shows bisection with 0.1-s and 0.9-s standards

examples above), and others finding near-geometric mean bisection (Allan & Gibbon, 1991). Arithmetic mean bisection seems compatible with the idea of "equal distance" applied to a linear and accurate time scale, as the duration value an equal linear distance in time from *S* and *L* is, of course, the arithmetic mean.

In fact, the controversy between near-arithmetic and near-geometric mean BP was revealed to be more apparent than real, but attempts to resolve this issue yielded new insights about bisection performance. Wearden and Ferrara (1995) investigated the effects of different spacing of comparison stimuli between constant *Short* and *Long* standard values. For example, one spacing could be arithmetic, with equal real-time spacing between consecutive values (e.g. 100-ms steps). On the other hand, spacing might be logarithmic, so the logarithms of *S* and *L* are taken, the difference between them divided into equal parts, and finally, converted back into time values. Logarithmic spacing produces a greater number of values below the arithmetic mean than above it; in other words, there are a larger number of shorter comparison durations than longer durations in logarithmic spacing. Wearden and Ferrara found that bisection points were lower in the logarithmic than the linear spacing conditions. In general, for any particular comparison duration value, a greater number of L responses occurred if this value was part of a logarithmic stimulus set as opposed to a linear one. Clearly, the dependence of the proportion of L responses on stimulus spacing, even when the *S* and *L* standards are the same, calls into question the idea that the bisection point is some sort of psychological mid-point between these standards.

Wearden and Ferrara's work went further. As outlined above, the normal method of conducting bisection experiments is to initially present *S* and *L*, identified specifically as standards. But suppose you do not do this. Wearden and Ferrara used a method that they called *partition bisection*. No explicit standards were presented, but the comparison durations occurred in random order, and the participants' task was to decide whether each should be classified as S or L, using any criterion they chose. Initially, individuals needed a few exposures to the stimuli to be judged, but a number of repetitions of the series of stimuli produced an orderly psychophysical function. In fact, statistically, the psychophysical function derived from partition bisection was no different from that obtained with "normal" bisection, when the *Short* and *Long* standards are explicitly presented at the start of the experiment. How can this be? Wearden and Ferrara supposed that people were not really using the initial standard values to make their decision about each comparison duration, but rather some measure of the central tendency of the durations of *all* the stimuli presented, whether they were standards or comparisons. Their model is discussed later in this chapter, and predicts the kind of

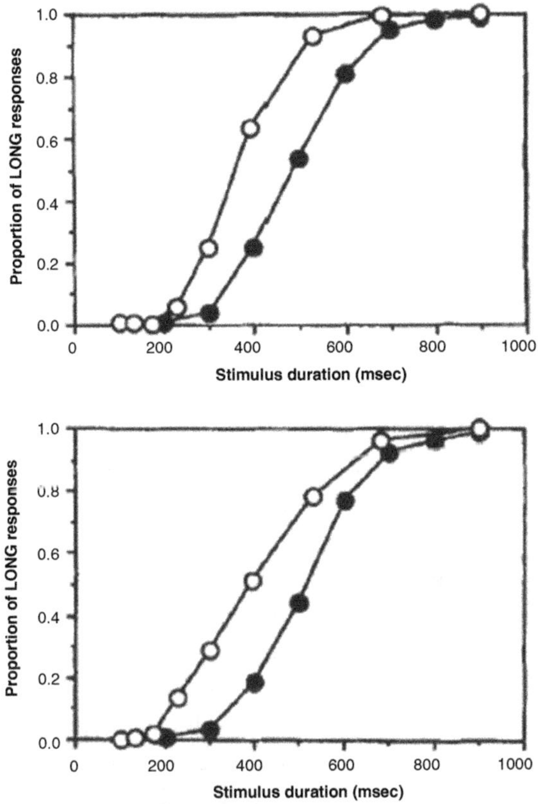

Fig. 4.3 Data from Wearden and Ferrara (1995). *Upper* and *lower panels* both show bisection performance with the *Short/Long* pair of 100/900 ms. In both panels, logarithmic spacing of comparisons is shown by open circles, and linear spacing by filled circles. *Upper panel*: Bisection performance with the "similarity" method (i.e. with presentation of identified *Short* and *Long* standards). *Lower panel*: Bisection performance with the "partition" method, without explicit presentations of standards

stimulus spacing effects that Wearden and Ferrara (1995) found. Although it may seem unlikely that people would give no special status to the *Short* and *Long* standards in bisection, other data support this view (e.g. Allan, 2002). Figure 4.3 shows data from Wearden and Ferrara (1995) and illustrates both the stimulus spacing effects and the similarity between the results obtained from normal bisection (with standard presentations)—using what Wearden and Ferrara called the *similarity* method—and partition bisection.

I turn now to a description of various formal models of bisection performance, some of which have been slightly simplified from the original version, with the intention of capturing the essence of the ideas that they embody.

Wearden (1991b)

The basis of the Wearden (1991b) model was the simple idea that bisection decisions were based on the *difference* between the duration of the stimulus to be classified, t, and S and L. Supposing that all durations on average are represented accurately, the bisection point would then be determined by the duration that was the same temporal distance from S and L, and so the value t would be that which satisfied $L - t = t - S$, producing $2t = (L + S)$ or $t = (L + S)/2$, that is, arithmetic mean bisection. If S and L were represented by Gaussian distributions, with accurate means and some coefficient of variation, c, with values S^* and L^* randomly sampled from these distributions on each trial, then the bisection point on average still remains at or very close to the arithmetic mean, although the psychophysical functions will flatten when c is larger.

However, data from Wearden (1991b) did not support exact arithmetic mean bisection: the bisection point was close to the arithmetic mean, but slightly below it. To account for this, I added what is in retrospect a sort of bias parameter, in a slight variant of the idea of simple differences which I called the "modified difference model." I assumed that people calculate the two absolute differences between t and S^* and L^*; let us call these $D(L^*,t)$ and $D(S^*,t)$. If these are sufficiently different, meaning that they differ more than some quantity x, then the bisection decision is based simply on discerning the smaller of the two differences. If the difference between $D(L^*,t)$ and $D(S^*,t)$ is less than x, the t value is ambiguous with respect to S and L, and in this case the model always chooses L as the response. In other words, ambiguous stimuli were always resolved as L, producing a bias towards the L response and thus a shift in the bisection point to some value below the arithmetic mean. If x was small, the bisection point was just below the arithmetic mean; if x was larger, the bisection point could be well below the arithmetic mean.

Of course, the *x* parameter was an arbitrary device to shift the bisection point towards the values where it was found to be experimentally, and has no real theoretical justification, except possibly to say that scalar variance causes representations of *L* to vary more than those of *S*, and thus to be more difficult to distinguish from some intermediate time, *t*. However, the model fitted data well, and was able to be further developed by adding other parameters, such as a "random responding," to deal with data from bisection experiments using young children in Droit-Volet and Wearden (2001) and later articles, as discussed in Chap. 7.

Allan and Gibbon (1991)

The bisection model used by Allan and Gibbon (1991) derives from a general mathematical treatment of bisection provided by Gibbon (1981). This is too complex for presentation here, but their article provides the results in simplified form. In summary, their model proposed that a long response occurred for some duration value, *t*, if

$$t^2 > \left(l^* s^*\right) / \beta \tag{4.4}$$

Here, s^* and l^* are samples from the memories of the *Short* and *Long* standards, and β is a bias towards the L response. I have changed their terminology so that it is consistent with that from the other bisection models discussed here.

The bisection point, *BP*, is given by

$$BP = \sqrt{}\,(sl) / \beta \tag{4.5}$$

that is, the bisection point is at the geometric mean of the *Short* and *Long* standards divided by the bias value. When fitted to data, their model involves two parameters: one is the bias value, β, and the other is γ, the coefficient of variation of the memory representations of the *Short* and *Long* standards. In fact, the bias value reported in their 1991 article was very close to 1.0, so in effect, the model predicted geometric mean bisection. However, it was able to fit data showing a bisection point at another location by changing the bias value. Although the basis on which it was constructed was quite different from that of the Wearden (1991b) model, there are clear similarities between the parameters derived from fitting the two models to data, as both involve a memory/timing variance parameter, and both involve some factor that potentially biases the model towards or away from making the L response after each comparison duration. In addition, both use the values of the *Short* and *Long* standards as part of their calculations.

Wearden and Ferrara (1995)

Both the model of Allan and Gibbon (1991) and that of Wearden (1991b) are of the form *decision* = *f*(*t*, *S*, *L*); that is, the decision as to whether *t* should be classified as S or L is some function, *f*, of the value of *t* and the values of *S* and *L*. The functions are different between the two models, but the principle is the same. This means that for both models, if *S* and *L* are constant and *t* is the same, the decision should then also be the same. The models are thus "context-free": the distribution of time intervals between *S* and *L* should have absolutely no bearing on the decision made about a particular *t* value. However, Wearden and Ferrara (1995) showed that the distribution of comparison values between constant *S* and *L* values *did* affect the psychophysical function. For example, psychophysical functions were shifted to the left if the spacing of comparisons between *S* and *L* was logarithmic rather than linear (see Fig. 4.3). A particular *t* value was more likely to be classified as L if it was a member of a set of logarithmic spacing of comparisons rather than a linear one, so the decision as to how to classify *t* is clearly dependent on something more than *t*, *S*, and *L*. Neither the model of Allan and Gibbon (1991) nor that of Wearden (1991b) can deal with this sort of "context-sensitive" effect, and so something else was needed.

Wearden and Ferrara's (1995) experimental procedure varied the spacing of comparisons between *S* and *L* in a number of ways. One variation was the comparison between logarithmic and linear spacing, already mentioned, but others involved artificial distributions in which there were a greater number of comparisons above the mean of *S* and *L* than below it, or vice versa, and these manipulations also shifted psychophysical functions. A review by Kopec and Brody (2010) of a large number of experiments on temporal bisection in humans revealed that this effect was quite general: "logarithmic-like" spacing shifted the bisection point to the left relative to linear spacing.

In addition, as mentioned earlier, Wearden and Ferrara developed *partition bisection*, where no explicit standards were presented. Partition bisection was found to produce psychophysical functions seemingly identical to those obtained with the "similarity" method involving initial presentations of *S* and *L* (see Fig. 4.2, for example). The fact that the *S* and *L* standards appeared to have no special status in bisection, allied with the effects of comparison spacing between fixed *S* and *L* values, led us to propose that individuals were classifying some comparison *t* not in terms of its relation to *S* and *L*, but to the *mean* of all the durations presented, regardless of whether they were identified as the standard. Fundamentally, if some *t* value were perceived as *shorter* than the mean, an S response would result; if *longer* than the mean, an L response would occur. If the *t* value was ambiguous, in that it differed from the mean by less than a certain threshold amount, the model responded S or L at random with equal probability.

More formally, if *t* is the duration to be classified, and *M* the mean of all the stimulus durations presented, the model responded L if

$$(t - M)/t > b \tag{4.6}$$

and S if $(t-M)/t< b$. In the condition where $b<(t-M)/t< b$, the model responded S or L at random with equal probability. To simulate trial-by-trial variance, *M* was represented as a Gaussian distribution with mean *M* and some coefficient of variation, *c*, so that on each trial, a value M^* randomly sampled from this distribution was used to represent *M*. The rule for determining the degree of similarity between *t* and *M* was the same as that used in the fixed-threshold model of temporal generalization developed by Wearden (1992), and thus Wearden and Ferrara's account of bisection promoted a degree of theoretical similarity between the treatment of bisection and temporal generalization.

This model easily dealt with the stimulus spacing effects obtained by Wearden and Ferrara. A logarithmic spacing of stimuli between *S* and *L* results in a mean of all the stimulus durations which is smaller than that derived from a linear spacing between the values. This lower mean results in any particular *t* value being more likely to be classified as L in the logarithmic case than in the linear case, as it is more likely to be above the mean, and the same idea accounts for the other stimulus spacing effects found by Wearden and Ferrara (1995). In addition, the model proposes no special role for the shortest and longest stimuli in the set, thus suggesting that psychophysical functions from partition bisection should be identical to those found in "normal" bisection, which is what Wearden and Ferrara (1995) found.

Although Wearden and Ferrara's model dealt well with stimulus spacing effects, and was compatible with results from partition bisection, it has attracted little or no interest since the time of its publication, other than a re-invention of something similar to its basic idea by Brown, McCormack, Smith, and Stewart (2005) in a model dealing with bisection of stimuli in other modalities as well as time.

Kopec and Brody (2010)

In their 2010 article, Kopec and Brody first provide a detailed review of the main results from a number of experiments on temporal bisection with human participants. I will mention here only the most important. Firstly, they identify the location of the bisection point in the data set as a whole as "sub-arithmetic": that is, the bisection point is located at some value below the arithmetic mean of the *S* and *L* standards. Given that, mathematically, the geometric mean of *S* and *L* is always below the arithmetic mean, this sub-arithmetic mean value can be close to the geometric mean, although Kopec and Brody regard their characterization of the bisection point location as more accurate overall, as cases exist where the bisection point can be below the geometric mean or very close to the arithmetic mean. Their review confirms two effects reported in Wearden and Ferrara (1995, 1996). The first is that the bisection point is lower when the spacing of comparison durations between *S* and *L* is logarithmic rather than linear, and the second is that the closer *S* and *L* are to each other (i.e. the lower the *L/S* ratio), the more sub-arithmetic the bisection point tends to be. In addition, their review confirms the finding reported by Ferrara, Lejeune, and Wearden (1997), among others, that small *L/S* ratios produce smaller Weber fractions (i.e. more sensitive timing) than larger ratios (although there are some data which do not show this; see Droit-Volet & Zélanti, 2013a).

Kopec and Brody account for the overall pattern of results, including some effects not mentioned here, by a two-step model, which I present in outline form in this section. The first step involves deciding whether each comparison stimulus is *S*, *L*, or some intermediate value. Stimulus durations close to *S* or *L* are classified as S or L responses in this initial stage. If the first-stage decision mechanism decides that a comparison duration is *neither S* nor *L*, a second step is taken. This involves comparing each comparison *t* with *S* and *L* using a modification of a similarity rule like that used in Wearden (1991b). The difference between *t* and the corresponding *S* and *L* values is subjected to a bias, which is based on the "gambler's fallacy," the belief that after a series of heads in a coin toss, for example, a tail becomes more likely. Here, if the model has made many L decisions, it becomes increasingly biased to respond S, and vice versa.

Specifically, the model responds S if $abs(S-t)*b<abs(L-t)$, and L otherwise, where b is a bias factor. The bias begins at 1 (i.e. no bias), and each S response increases the bias by an amount x, whereas each L response decreases it. The model has two parameters. One is the coefficient of variation of the representations of *S* and *L* in memory, and the other is the rate (x) at which the bias changes across trials.

The model can account for a range of effects obtained (linear versus logarithmic spacing, for example), and is perhaps the most comprehensive yet produced. It is not, however, without its problems. The standards *S* and *L* have a special status in the model, and in fact their memory distributions are central to its operation for both the first and second steps. This poses potential difficulties in simulating partition bisection where *S* and *L* are not presented separately at the start of the experiment, and thus presumably not identified by the participant in any special way. As Wearden and Ferrara (1995) showed, partition bisection can produce results seemingly identical to those obtained when *S* and *L* are specially identified (Fig. 4.2). Overall, however, despite this and some other difficulties, the Kopec and Brody model probably provides the best current account of the general pattern of data obtained in bisection experiments with humans.

Although there are clearly a number of competing models, especially models of bisection, the models have some common features overall. One is the presence of some parameter representing variability in memory/timing. The other is some decision process, or bias, which influences the number of responses of a particular type (YES/NO in generalization, S and L in bisection). The models of bisection do differ, however, with respect to the basis of the classification decision: most models assume that the *S* and *L* standards are the basis for some sort of comparison, although the Wearden and Ferrara (1995) model does not assume this. As I point out elsewhere (Wearden, 2004), there is a degree of theoretical continuity between some models of bisection and generalization. The Wearden and Ferrara (1995) model of bisection, as well as the MCG model [Eq. (4.2)] and the episodic generalization model [Eq. (4.3)], can all be classified as models involving "thresholded normalized differences." In other words, the difference between two durations (one typically a comparison and the other some sort of standard) is divided ("normalized") by some quantity, and the result is compared with a threshold. If it is below the threshold, one response results, and if it is not, another response is produced. A degree of theoretical consistency in the treatment of different types of generalization and bisection seems intuitively reasonable, as both types of tasks involve comparison between a duration which is to be classified and some sort of standard or standards.

In practice, the models most frequently applied to data have been variants of the MCG model of temporal generalization and of the Wearden (1991b) model of bisection, as discussed in Chaps. 7 and 8. The latter is probably not the "best" model of bisection in terms of fit to the range of data (Kopec and Brody's model probably performs best overall), but from an application perspective, this may not be so important. For example, if we are comparing the performance of different groups of children (as in Droit-Volet & Wearden, 2001, for example), then one difference obtained (discussed in more detail in Chap. 7) is the changes in memory/timing variance as children age. The same conclusion almost certainly would have been drawn had different models of bisection been used, as in order to account for psychophysical functions which differ in steepness as children age, the memory/timing parameter would almost certainly have shown the same ordinal changes irrespective of the bisection model used. This suggests that models can be useful even if they are not the best in terms of fit to the overall pattern of data obtained.

Summary

The material in this chapter has introduced the reader to some of the background and simple mechanics (but not all of the details) of models of temporal generalization and bisection, both very commonly used tasks, consistent with SET. The models generally fit data well in terms of goodness of fit, at least by normal standards in psychology, and some of them also "behave" correctly when experimental conditions such as stimulus spacing are changed. Most of the models have a parameter representing timing/memory variability, and another derived from decision processes. In temporal generalization models, this determines the overall level of YES responses (decisions that a comparison stimulus has the same duration as the standard), and in bisection models, the leftward or rightward shift of the psychophysical function, as reflected in the proportion of S and L responses to each comparison duration. The ways in which such models are applied in detail to data from different participant groups is explored further in Chaps. 7 and 8.

5

Cognitive Processes, Emotion, and Timing

In this chapter, I discuss work on cognitive processes and timing, and review some fairly recent research on emotion and timing as well. Cognitive processes and timing have been linked in various ways in the literature. Multi-process models of timing like SET, for example, with its tripartite division into clock memory and decision processes, embed the basic timing mechanism in a more complex cognitive framework, what Treisman (2013) called the "clockwork" of the timing system, as opposed to the clock itself. The link is even more evident in recent multi-process models such as that of Taatgen, van Rijn, and Anderson (2007), where a timing mechanism is used as a "front end" to the ACT-R system (Anderson et al., 2004), which has been widely used to account for performance on many non-timing tasks.

Historically, another link between a "general" cognitive process and timing comes via the study of attention and timing. If we consider perceptual tasks which do not involve timing, such as discriminating colours, the concept of attention is usually not invoked in simple circumstances. For example, suppose we simply present individuals with red or green squares, and they must merely say which is which. All individuals with normal colour vision will be 100 % accurate in this case. We do not,

J. Wearden, *The Psychology of Time Perception*,
DOI 10.1057/978-1-137-40883-9_5

however, normally say that they are "attending to colour" (although they must be), and invoking the idea of "attention" in the basic process of colour discrimination seems to add nothing. Of course, there are more complex cases where individuals might be asked to attend to, and report, the colour of a stimulus rather than its shape, location, intensity, or other variables, and using the idea of attention here seems more natural, even though it would not be used in the simple case.

In timing research, however, some theorists argue that "attention to time" is an essential part of time perception itself, even in the simplest cases, such as judging the duration of a tone without any competing stimuli (e.g. Zakay, 1993). Why is this? One prevalent theory proposes that an inner timing process, usually identified with something akin to SET's internal clock, needs attention to function at all. For example, if we use SET's clock as an example, it may be that the switch connecting the pacemaker to the accumulator will remain closed—thus allowing ticks to flow from one to the other—only if attention is paid to time. If it is not, in the case where attention is diverted away from the processing of time to some other task, the switch then opens, pulses are missed, and the time judgement is affected, as diagrammed in Fig. 5.1. The concept of attention therefore seems to play a much more fundamental role in time perception in simple circumstances than in the perception of colour, for example.

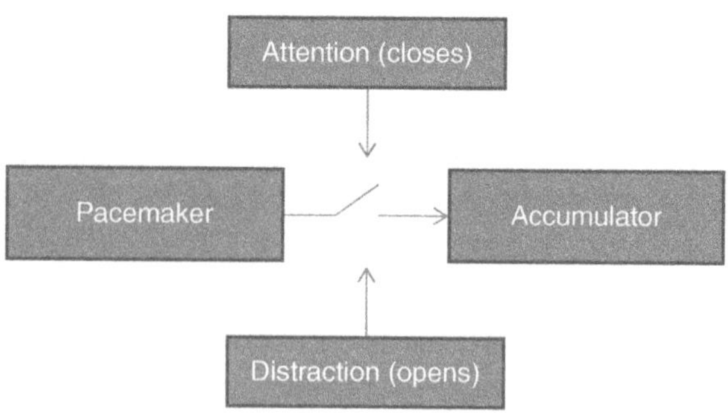

Fig. 5.1 Attentional effects on timing. The pacemaker is connected to the accumulator via a switch. The switch is closed when attention is paid to time, but opened when attention is diverted away from time

A frequently cited article by Thomas and Weaver (1975) is often used to justify the position that attention to time is central to time perception, and that participants' judgements are based on the "amount of attention" paid to temporal features of a stimulus relative to its non-temporal features. Thomas and Weaver's study does have unusual aspects, however, which render its links to more modern ideas regarding attention and time perhaps questionable. In their study, participants were initially trained until they were reliably able to associate very short presentations of a blank visual field (20, 60, and 100 ms) with "short," "medium," and "long" responses (although performance figures are not given in the article). They were then presented with three different sorts of stimuli, each either 40 or 80 ms long. The stimuli were blank fields, words consisting of three letters, or the letters of these words scrambled, and participants used the classification scheme previously learned. Perceived durations were longer for both the 40-ms and 80-ms stimuli (which were clearly distinguished) when letters were presented than for the blank fields, regardless of whether the letters made up a word. Thomas and Weaver concluded that the duration judgement was based on the output of both a temporal processor and one that processed visual information. Rather than competing, however, these processors were combined in some way: recall that letters, presumably involving the contribution of both processes, were judged as longer than blank fields. The result and conclusion that temporal and non-temporal processing were added together when duration judgements were made contrasts with more modern ideas that temporal and non-temporal information processing tend to compete.

Many studies have attempted to observe the results of diverting attention away from a timing task, usually to one not involving timing, as in the "dual-task" procedures discussed below. As will be shown later, this sort of procedure involves more than just diversion of attention: attention must be diverted towards "something else," and the nature of the "something else" has been found to be important in determining how, or whether, the timing task is affected by the diversion of attention. This suggests the existence of some competition between timing tasks and other sorts of cognitive tasks; for example, they may all depend on a common pool of limited resources. In the first section of this chapter, I will review some experiments involving manipulations of attention to time. However, there is another way of examining potential relations between cognitive processes and timing more generally, and this is to look at correlations between performance on some cognitive task and aspects of performance on a timing task. For example, are people who are better at, say, working memory tasks also "better timers" because they show greater sensitivity to duration than those who are poorer at memory tasks? I will call this type of research "predictive studies": here, the aim is to determine whether performance on a timing task is predicted by scores on some other task.

Attention and Timing

In studies of attention and timing after Thomas and Weaver (1975), most research has been based on the premise that timing and non-timing processes compete, such that competing non-temporal information causes duration judgements to be shorter than those without the competition. The most common experimental arrangement for investigating this phenomenon is the so-called dual-task procedure. There are various iterations of these, but one of the simplest cases involves the interposition of a single non-timing task during the timing of an interval, as in Fortin and Rousseau (1987). The timing task was the production of a 2-s interval, and participants received initial training to produce fairly accurate responses. The test condition involved concurrent performance of a memory search task, similar to that of Sternberg (1966). Before the production started, participants were presented with a set comprising one to six digits (i.e. the "set size" was one to six digits), containing values from 0 to 9. The interval was then started with a press on the middle button of three available response possibilities. After 500 ms, a digit was presented, and if it was in the set shown before the interval started, participants were instructed to terminate the interval with a right button-press, and if it was not in the original set, with a left button-press. The interval produced lengthened systematically with increasing set size, and was longer when the digit was not in the set than when it was.

The sketch in Fig. 5.1 can easily be used to account for the basic effect of response lengthening. When the digit is presented during the interval, attention is diverted away from timing for a period needed to perform the non-timing task (which is longer when the set size is larger, as in Sternberg, 1966). During this time, the switch opens, and pulses from the pacemaker are missed, so the interval produced is lengthened, as the time needed to accumulate the number of pulses the participant previously learned as a representation of 2 s is increased.

Fortin, Rousseau, Bourque, and Kirouac (1993) conducted a variant of the Fortin and Rousseau study, sometimes using different non-timing tasks such as visual search. In their Experiments 3 and 4, the participant was required to search for targets in a display containing non-target distractor items. Targets were either a circle with a vertical line in it (contrasted with plain circles) or a green letter T, with blue letters T and X and a brown or green S as potential distractors. As the number of distractors increased, the length of time to find the target item when it was present increased. However, the production of 2 s remained constant (at a value around 3.2 s) as the number of distractors changed. Obviously, this suggests that the interference between timing and the non-timing task is dependent on the cognitive processes used for the non-timing task. When short-term memory was loaded, as in the Fortin and Rousseau study, higher loads produced increasingly long interval productions, but visual search that appeared to require no short-term memory produced an increase in production over that found when there was no second task, although the production time did not increase as the difficulty of the visual search increased.

The situation was further complicated by results from Fortin and Breton (1995). They again used the Sternberg task, but this time it was the set of potential items that was presented *during* the interval production. The target stimulus about which a decision had to be made was presented *after* the interval had been produced. Now the set size had no affect on the interval production, although a reaction-time task showed the usual effect in that the time taken to judge whether the target item was or was not in the set increased with set size. Other experiments in this article showed that the number of syllables in words involved in a rhyme judgement task or the degree with which complex figures had to be mentally rotated all increased interval production duration. With respect to short-term memory tasks, Fortin and Breton concluded that the critical factor in lengthening the interval production was the time taken to decide whether items were in the set, rather than simply the load imposed by the set size itself, although other sorts of tasks not involving short-term memory obviously also lengthened productions.

Using a procedure similar to that in Fortin and Rousseau (1987), Fortin, Duchet, and Rousseau (1996) presented target items between 17 successive taps, each 2 s apart, and obtained a lengthening of the inter-tap interval with an increase in set size. Another study by Fortin and Rousseau (1998) looked at set-size effects in reproduction. The interval to be reproduced was defined by two tones, and the participant reproduced this interval by making two taps. A set of digits was presented, as in the Sternberg (1966) task, and then a target item. In one condition, the target item was presented during the reproduction phase but not during the example of the interval to be reproduced. The reproduction tended to lengthen with set size, although the effect was not large. In contrast, if the target item was presented between the tones defining the interval to be reproduced, but not the reproduction, the reproduced time shortened with set size. Both these effects, of course, are consistent with the notion that the memory search task detracts from temporal accumulation.

Overall, then, studies with a single item presented during the production or reproduction of an interval are broadly consistent with the description of attentional effects shown in Fig. 5.1, where the non-timing task might require a single "interruption" in the flow of pulses from the pacemaker to the accumulator, although it remains rather unclear just exactly what processes interfere with timing, and why.

Other procedures which manipulate attention to time fit less well with the single-interruption idea, and involve what seems to be continuous "sharing" of attention between two tasks rather than a single "excursion" away from the timing of an interval while a non-timing task is carried out. One of the best-known series of such experiments is that of Brown (1997). Here, the timing task was the production of a continuous sequence of 2-s or 5-s time intervals, during which participants were required to perform a non-timing task. In different experiments, this was a pursuit rotor task, visual search, or mental arithmetic, and each task had two levels of difficulty. Performance on the timing task showed consistent effects. The productions were lengthened (indicating "loss of ticks," as in the cases discussed above), and became more variable regardless of the non-timing task used, whereas the more difficult versions of the non-timing task tended to have a greater effect on the task. However, Brown was also interested in the issue of "bi-directional interference": if the non-timing task disrupted the timing task to some extent, was the effect also found in the other direction, with performance on the non-timing task deteriorating when a timing task was performed simultaneously? Here, the results were more complex. Performance on the pursuit rotor task and visual search showed no deterioration when the timing task was performed concurrently: in fact, performance on the pursuit rotor apparatus improved. However, deterioration in mental arithmetic performance was observed when timing was carried out simultaneously.

One theory often proposed to explain the interference effects between cognitive tasks in general is that the tasks compete for some common pool of resources. Tasks which interfere strongly with one another presumably compete for very similar resources, whereas those which do not interfere at all are perhaps performed by different cognitive systems. The lack of bi-directional interference in some cases also suggests that, although there is competition between a timing and non-timing task for some resource, this resource is used more for timing (so timing is interfered with) than for the non-timing task in question.

Another type of procedure which appears to involve some sort of continuous attentional sharing was developed by Macar, Grondin, and Casini (1994). This method required participants to divide their attention between temporal and non-temporal aspects of a stimulus, with instructions indicating the degree of attention that should be paid to each. For example, in their Experiment 1, words were presented, and the task was (a) to count animal names and (b) to reproduce the duration of the word just presented by pressing a button. Control conditions requiring 100 % attention to either the duration or the words were included, but attention was also manipulated more finely: 25 % words, 75 % duration; 50 % to each; or 75 % words, 25 % duration. The durations were reproduced as progressively shorter compared to the 100 % duration condition as the attention paid to time decreased. Two other experiments used a similar method, but here the non-timing dimension was the intensity of either an auditory or visual stimulus. In general, paying less attention to duration shortened perceived duration. Although it is difficult to ascertain exactly how the participants in this study were conforming to general task instructions (and even more difficult to know whether they were allocating their attention accurately), the effects on timing performance were consistent and aligned well with the general idea that "attention to time" is an essential feature of time perception when prospective timing procedures are used.

Brown, Collier, and Night (2013) conducted a complex series of dual-task experiments, this time involving competing tasks which tapped into executive functions. As the authors note, some theorists regard executive processes as linked to "general" fluid intelligence or working memory. An influential article by Miyake et al. (2000), however, postulated that executive processes involved three specific and separate components. These were defined as *shifting* (also known as attentional switching or task switching), which "involves shifting attention back and forth between multiple tasks or cognitive operations"; *updating*, "which involves keeping track of old and new information"; and *inhibition*, "a basic executive control processes that involves suppressing dominant, automatic responses and resisting distraction" (all from Brown et al., p. 949). The authors used a methodology which was a combination of traditional dual-task methodology and the "sharing" of processing manipulation described by Macar et al. (1994). Participants performed the basic timing task alone, which involved repeated production of an interval they judged to be 5 s. They also performed an executive task alone, and among different experiments this involved switching, updating, or inhibition. In addition, in other conditions, they were asked to divide processing resources 25/75 % timing/executive, 50/50 % timing/executive, or 75/25 % timing/executive.

In Experiment 1 (shifting), the non-timing tasks involved switching between global and local processing of letters. Large letters could be composed of small-letter components, which might be different from the large letters, and the task required individuals to switch between types of processing. Compared with the condition without the switching task, the productions of 5 s with the switch were longer and more variable, and productions lengthened as less attention was paid to the timing task. Furthermore, the interference between timing and switching was mutual: the time needed to complete the switch task was greater when it was accompanied by the timing task than when performed alone. In Experiment 2, an updating task was used. Squares, circles, or triangles were presented in random order, and the participants' task was to respond to YES or NO probes, with a response of YES if a particular shape had occurred four times. Each YES response reset the count requirement back to zero, so the task essentially involved updating the number of times each shape had occurred. Once again, the putative 5-s interval produced was lengthened and was more variable when a secondary task was performed, and once again, the effect was most pronounced when the smallest amount of time processing (25 %) occurred. However, evidence for bi-directional interference was weaker than in Experiment 1, with the only marked effect of the concurrent timing task when attention to the updating task was at its smallest (25 %), in which case the response time was lengthened. Experiment 3 (inhibition) used the well-known Stroop interference effect. Participants were tasked with identifying words corresponding to colours using four responses on a response box (red, green, blue, yellow). The words corresponding to these colours were presented, but in half of the trials the colour used to display the word differed from the colour word itself (e.g. the word GREEN written in red). For the other half of the presentations, the word and its displayed colour were congruent (e.g. GREEN written in green). As in the other two experiments, the target 5-s interval was produced as longer (and more variable) with the secondary task than without, and the lengthening was greatest when the smallest amount of attention was paid to time. In contrast to Experiment 2, the timing task interfered markedly with performance of the executive task, and the amount of attention paid to the executive task also had an effect.

Overall, then, Brown et al.'s study showed that performing all three executive processes (shifting, updating, and inhibition) perturbed timing performance by lengthening the intervals produced and increasing their variability. In addition, the attentional instruction (similar to that in Macar et al., 1994) intensified the effect of the secondary task. Performing the timing task interfered with performance on the switching and inhibition tasks, but less so on the updating task, which suggests that interval production competed for cognitive resources with switching and inhibition but not with updating. Another investigation of this question using a different methodology is the study of Ogden, Wearden, and Montgomery (2014), discussed in the next section.

Many of the studies presented above showed that imposition of non-timing tasks on top of a timing task resulted in shorter judged durations, either by diverting attention away from timing or by some other interfering operation. But is the reverse true? Does increasing attention to time cause stimuli to be judged as longer? One difficulty in answering this question is that in most experiments where participants of student age are judging the duration of brief stimuli, one might expect that attention is at its maximum. To address this question, Mattes and Ulrich (1998) devised a method that they called "directed attention." Participants received cues with information about stimuli to be presented, but the cues were not always valid, and the participants were aware of this. For example, a presented stimulus could be either auditory or visual, and a cue with validity 0.5, 0.7, or 0.9 preceded the stimulus. In the case of 0.9, the cue gave correct information as to modality on 90 % of trials and incorrect information on 10 %. In other experiments, it was the location of a stimulus in the visual field (right or left) that was cued, again sometimes incorrectly. Different methods were used among the different experiments, but in general, increasing the probability of correct cuing increased perceived duration. However, in none of the experiments was a 100 % correct condition used, and so we do not know whether the duration at, say, 90 % correct was shorter or longer than it would have been at 100 % correct, although trends in data suggest that it would have been shorter.

Another situation in which increasing attention appears to increase subjective duration is in the *oddball* paradigm, where the duration of a stimulus seems longer if it is unexpectedly different in some way from the rest of the sequence of stimuli in which it is embedded. Tse, Intriligator, Rovest, and Cavanagh (2004) looked at what they called "time's subjective expansion" when oddball stimuli were presented. In their Experiment 1B, for example, the standard stimuli were fixed black circles, and the oddball was an expanding black circle. The oddball stimulus was judged as longer (on average about 25 % longer) at all stimulus durations (which varied from 75 to 2100 ms) except the shortest, where the oddball stimulus was rated as shorter than the standard. The authors conducted a number of other experiments with similar results but which investigated conditions affecting the oddball effect, and concluded that, in general, the relative increase in attention to the oddball stimulus increased its subjective duration. For further discussion of the processes involved in the oddball effect, see Birngruber, Schröter, and Ulrich (2014).

Theoretical Explanations of Attentional Effects

The material discussed above clearly supports the view that diverting attention away from timing towards some non-timing temporal processing typically shortens time judgements, at least with the methods used. It is less clear whether increasing attention to time has the opposite effect, although results from the oddball procedure suggest that this is possible.

How can the general pattern of results be explained, and more specifically, where is the locus of attentional effects in formal models of timing? Lejeune (1998) provides a thoughtful review of how the concept of attention has been used in timing in both animals and humans, and I will follow some of her discussion here. Suppose that we have a timekeeping process whereby some kind of "units" of time are generated and stored, such that the number of units represents duration. Such a model can be identified with the internal clock of SET, but may be more general. One possible model of attentional effects on timing is to suppose that attention is needed to maintain the connection between the generator and the store. If attention is diverted away from timing, then the connection is severed, and no time units are stored until the connection is restored (as in the sketch in Fig. 5.1). The switch of the internal clock proposed by SET operates in this way: it is either open (and the pacemaker-accumulator connection is severed) or closed (and clock ticks are accumulated); there is no intermediate state of partial closure. Such a simple model fares best in accounting for data like those of Fortin and Rousseau (1987). A single non-timing probe is presented while a timing task is in progress, and thus it is easy to imagine that the switch opens when the non-timing task starts, attention is diverted to the non-timing task, no temporal accumulation occurs while the switch is open, and the switch is then closed again, and pulses continue to accumulate until the timing task trial ends. In this case, it is obvious that, all else being equal, fewer time units will be stored in the dual-task condition than in that without the non-timing task, and attention can keep the switch closed throughout the timing task.

However, this model can encounter difficulties even in simple cases. It assumes that the time for which the switch is open (and accumulation is stopped) reflects the time needed to perform the non-timing task, so less temporal accumulation will occur when more time is needed to complete the non-timing task—because of greater difficulty, for example—and thus there should be a relation between the shortening of judged time and the processing time of the non-timing task. As the work by Fortin and colleagues shows, this is sometimes the case, but it is by no means always found. The simple model at first sight also appears less able to fit to data from situations in which both the timing and non-timing tasks are continuous (e.g. the pursuit rotor task and repeated interval production concurrently, as in Brown, 1997) or attention is divided by instructions (as in Macar et al., 1994). Both situations seem to involve some sort of *sharing* of attentional resources between timing and non-timing tasks carried out simultaneously, rather than the "all-or-none" operation of the switch described above.

A better fit in these cases may be achieved with the "attentional gate model (AGM)" of Zakay and Block (1995), diagrammed in Fig. 5.2. Here, the path between the pacemaker and accumulator not only has a switch, but also has an *attentional gate*. The switch is supposed to operate automatically: it closes when the event to be timed begins, allowing temporal accumulation, and opens again when the timed event ends. In the AGM model, however, the attentional gate, not the switch, is the locus of attentional effects. Moreover, this gate does not operate in an all-or-none fashion, but instead produces graded effects: when the gate is open wide, pulses flow easily from the pacemaker to the accumulator, and when it is narrower, the flow is reduced. Because a task like that of Macar et al. or Brown is accounted for by different widths of the attentional gate, with instructions to pay attention to time affecting gate width, this graded effect of attentional allocation should produce a graded effect on time judgements, and it does, at least according to the results of Macar et al. (1994).

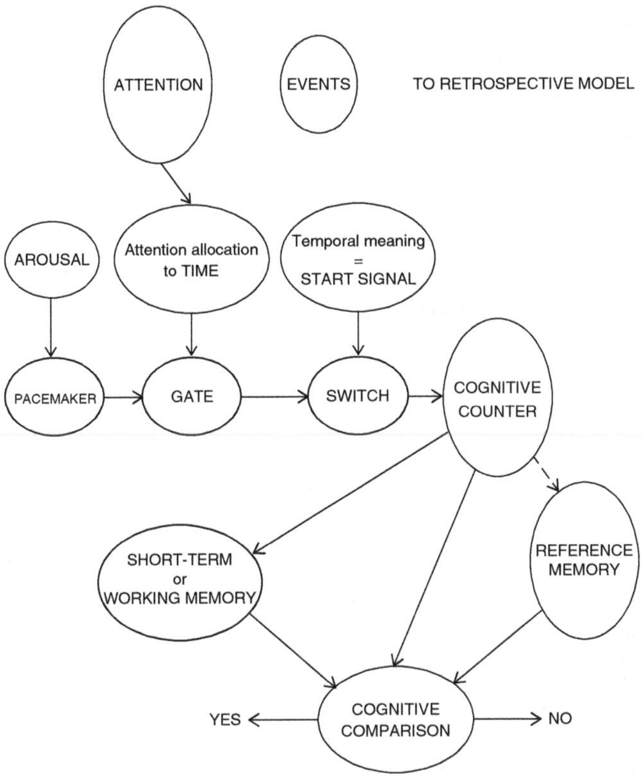

Fig. 5.2 Attentional gate model, after Lejeune (1998). An arousal-sensitive pacemaker send ticks to an attentional gate, the width of which is governed by attention to time. The ticks are then sent to a switch which is automatically controlled by stimulus onset and offset. When the switch closes, the pulses flow to the accumulator. The attentional gate varies the rate of ticks that pass through it in a continuously graded way

Lejeune (1998) argues that, in practice, distinguishing between the AGM and the switch model may be more difficult than it first seems. In the case discussed above, we assumed that the switch opens and closes once during the timing task, while the non-timing task is being performed, but this may not be the case. Suppose instead that a "flickering switch" is proposed, where the switch can open and close repeatedly while a timing task is under way, and during each of these "excursions," some aspect of the continuously present non-timing task is carried out.

The flickering switch operates in a manner similar to a normal single-processor computer: the operating system gives the impression of sharing resources between operations, but the effect is typically produced by very rapid switching between one operation and another.

The simple mathematics of internal clock models, which were discussed earlier in this book (appendix to Chap. 3), show that changes in pacemaker speed, usually described as "arousal" effects (see Burle & Casini, 2001), produce graded, multiplicative effects on time judgements, such that different pacemaker speeds between conditions will have more obvious effects at longer intervals than shorter ones. In contrast, a single opening and closing of the switch in order to carry out a non-timing task will result in the loss of the same number of pulses regardless of the duration timed, as was previously discussed. However, both the attentional gate model and a rapidly flickering switch will produce graded effects, the former by variation in gate width, the latter by the number of flickers (which will increase as the interval timed lengthens). This means that both are difficult to distinguish from a simple change in pacemaker rate: in fact, mathematically, they will produce the same effect.

The notion of attention causing changes in pacemaker speed has not been popular, probably because arousal and attentional effects are thought to be different. It is possible, however, that data coming from "oddball" experiments (Tse et al., 2004) require exactly this. Another potentially problematic result with respect to distinctions between attentional and arousal effects comes from Wearden, O'Rourke, Matchwick, Zhang, and Maeers (2010), where participants were required to switch rapidly from a non-timing task to one involving timing. In the first two experiments of this study, individuals were required to estimate the duration of a tone (Experiment 1) or a square of colour on a screen (Experiment 2). Sometimes this was the only task to be performed, but on other trials a series of numbers in rapid succession were shown just before the tone or square, and participants were required to add them up. This "paced serial addition" was a difficult task for participants, particularly when the numbers were larger (from 11 to 19, for example, rather than from 1 to 9). How did this affect performance on the duration estimation task? In general, estimates were shorter after the addition than without it, but the effect was greater when longer rather than shorter time intervals were presented; in other words, it was a multiplicative "slope effect," generally thought to be indicative of changes in arousal. However, task switching is normally considered to be a manipulation which affects attention (Monsell, 2003) or which involves executive processes (see Ogden et al., 2014). Accordingly, before the experiment was conducted, I had anticipated an effect based on a "late start": for example, if you arrive at the cinema 10 min after the start of a film, you miss 10 min of the film, regardless of its length (an "intercept effect"). However, the results were different, as though people missed the same proportion of the film each time—if it was longer, they missed more minutes of the film—something which obviously fails to fit the "late start" idea. On the other hand, the results are compatible with a change in the width of the attentional gate, which remains constant for the entire duration of the stimulus, and so the gate is narrowed on trials where participants switch between the addition and timing tasks. The results are equally compatible, however, with a decrease in pacemaker speed after the switch in comparison with no switch. Unfortunately, there seems to be no obvious way to distinguish the AGM's predictions from those based on changes in pacemaker speed.

Predictive Studies

This section reviews research attempting to link performance on various sorts of timing tasks to performance on other cognitive tasks. This is accomplished not by trying to interfere with timing using a dual-task methodology, but by correlating aspects of performance on timing tasks with other measures.

One such study is that of Rammsayer and Brandler (2007). A number of temporal tasks were employed: threshold determination around 50 ms for filled and unfilled auditory intervals, threshold around 1000 ms for filled auditory intervals, temporal generalization with standards of 75 or 1000 ms, a rhythm perception and temporal order judgement task plus determination of flicker fusion threshold, and a reaction-time task. Participants also received a number of measures of "primary mental abilities," including measures of verbal comprehension, word fluency, and perceptual speed, as well as non-verbal intelligence measures. The mental ability tests were used to produce a measure of "psychometric g," that is, general intelligence, and the principal finding of the article was the high correlation between psychometric g and "temporal g," a measure of the acuity of temporal discrimination derived from the timing tasks used. Rammsayer and Brandler were interested primarily in precision of temporal judgement as a predictor of intelligence, but the correlations obtained can also be interpreted in the reverse direction: higher general intelligence leads to higher temporal acuity. In a similar vein, Troche and Rammsayer (2009) investigated correlations between what they called the "temporal resolution power (TRP)" of an individual—derived from measures of temporal discrimination, temporal order judgements, and temporal generalization—and measures of psychometric intelligence. Their results suggest that TRP predicted psychometric intelligence well, but their path model showed that the relation was mediated by working memory: "higher TRP leads to better coordinated mental operations which, in turn, result in higher psychometric intelligence" (p. 479).

The fact that individuals with higher intelligence are better at discriminating stimuli is perhaps itself not surprising, as they may be generally expected to perform better on most cognitive tasks. Troche, Wagner, Voelke, Roebers, and Rammsayer (2014), however, sought to clarify the relationship between general discrimination ability, working memory capacity, and general intelligence, which they did by employing auditory and visual temporal discrimination as well as line length and pitch discrimination, and by relating these to both working memory capacity and general intelligence. Using various types of path models, they concluded that working memory capacity plays a central role in mediating the relation between sensory acuity and general intelligence, confirming the position of Troche and Rammsayer (2009).

The work just reviewed used only adult participants, but a series of studies by Droit-Volet and Zélanti not only tried to link aspects of performance on timing tasks to specific cognitive abilities assessed by psychological tests (many of them commonly used in neuropsychology), but did so in a developmental context, employing a participant population including both adults and children. More general questions about timing in children are treated in Chap. 7, but the Droit-Volet and Zélanti studies are included in this section as examples of correlational/predictive studies of aspects of time perception.

Given that memory and attention capacities develop with age, it seems feasible that any difference in performance on timing tasks among persons of different ages is, at least in part, due to developmental changes in these psychological functions. Zélanti and Droit-Volet (2011) investigated this explicitly. Children 5 and 9 years of age, along with adults, received four different bisection tasks carried out over 4 days. The stimulus for which duration was to be judged was a red circle. The four bisection tasks varied time values over a large range, with *Short/Long* pairs of 0.5/1, 1.25/2.5, 4/8, and 15/30 s. At all four duration ranges, the steepness of the psychophysical function (as assessed by its Weber ratio) increased systematically with age, the normal result found in other studies (see Chap. 7). Participants also received various neuropsychological tests assessing short-term memory (forward digit span) and working memory (backward digit span). There were tests of auditory attention as well, and an "attention/concentration" test derived from Cohen (1997). Adults not only showed better temporal discrimination than the younger children, but also showed better performance on the cognitive tests; therefore, the psychological test measures were converted to *z* scores for each age group, in which each individual score had the group mean subtracted from it, and this difference was divided by the group standard deviation. In other words, the psychological test scores were "normalized" by the average performance of the group. The *z* scores from the different age groups were combined to produce a final sample. Time sensitivity at the shortest duration range (0.5/1 s) was correlated with short-term memory span; in contrast, time sensitivity at the longest durations (15/30 s) was predicted by the attention/concentration score. For the intermediate durations (1.25/2.5, and 4/8 s), only age was a significant predictor of timing sensitivity.

The study just described kept the *Long/Short* standard ratio in bisection constant at 2:1, but varied the absolute values of the durations. A later study (Droit-Volet & Zélanti 2013a) varied not only duration range, but also the *Long/Short* ratio. As noted in Chap. 4, the *Long/Short* ratio in bisection is commonly regarded as a measure of the difficulty of the temporal discrimination that the task involves, with small ratios giving rise to more difficult tasks.roit-Volet and Zélanti (2013a, 2013b) studied children 5 and 8 years of age, along with adults, on bisection tasks with

Long/Short ratios of 6/5, 3/2, and 2/1. The stimulus for which duration was to be judged was a red circle. Furthermore, each ratio was tested at two duration ranges, one with all intervals less than 1 s, the other with all intervals greater than 4 s. Perhaps unsurprisingly, all participant groups struggled with the most difficult bisection range (6/5), and the authors used a measure of fit to individual psychophysical functions across conditions to ascertain the number of individuals who were successful on the task. In general, success improved in all groups as the duration ratio increased and with increasing age, but there were problems specifically with the 6/5 ratio: here, only 10 % of 5 year-olds, no 8 year-olds, and only around 40 % of adults succeeded at the short duration range. For the long duration range, the corresponding values were 0, about 25 %, and about 75 %. Comparison of Weber ratios showed that the adults produced much smaller values (i.e. more sensitive timing) than the two groups of children, who showed little difference as a function of age. Participants also received tests of short-term memory, working memory, and visual attention. Timing sensitivity as measured by the Weber ratio correlated negatively with age and with visual attention scores (thus higher visual attention scores were associated with more sensitive timing), but with neither of the memory measures, even though all test measures varied markedly with age. As in the other studies, *z* scores for the psychological tests were entered into the correlations between test scores and timing measures.

Zélanti and Droit-Volet (2012) examined bisection of stimuli in the auditory and visual modalities, with a 500-Hz tone as the auditory stimulus and the red circle used in the above-referenced studies as the visual stimulus. Two duration ranges were used, 0.5/1 s and 4/8 s, and the performance of children 5 and 8 years of age was contrasted with that of student-age adults. Various neuropsychological tests of memory and attention were also administered, including the Corsi block-tapping test, which is considered a measure of working memory capacity in the visual modality. Temporal sensitivity in both modalities increased with increasing age, but the age differences were particularly marked for the visual stimulus. Age was a significant predictor of temporal sensitivity in all cases, but another significant predictor at the longer duration range was the span of visual working memory, even when the stimuli presented were auditory.

Droit-Volet and Zélanti (2013b) carried out another study of temporal bisection using a blue circle as the stimulus, with *Short/Long* standard pairs of 0.5/1 and 4/8 s, and children 5 and 8 years of age contrasted with adults. A number of neuropsychological tests were administered, including a measure of information-processing speed derived from the Wechsler Intelligence Scale appropriate for the age of the participant (Wechsler, 1998). The typical result of increasing sensitivity to time with increasing age was found for both duration ranges, but the novel feature of this study was that the best predictor of timing sensitivity, apart from age itself, was speed of information processing. This increased dramatically with age, and correlated negatively with the Weber ratio. Once again, *z* score measures were used.

Overall, therefore, the studies by Droit-Volet and Zélanti showed that timing performance on bisection tasks can be correlated with differences in more general cognitive processes. Considering the studies as a whole, however, it is not completely clear just *which* cognitive processes are implicated. Sometimes scores on memory tests appear predictive (as in Troche et al., 2014), whereas at other times visual attention capacity seems to be involved or information-processing speed more generally.

The section above discussed the experiment by Brown et al. (2013), which used dual-task methodology with the non-timing tasks tapping three aspects of executive function. There is a predictive study with a similar aim. Ogden et al. (2014), like Brown et al., investigated *updating*, *switching* (Brown et al.'s *shifting*), and *inhibition*, but they also added *access* to semantic memory, another potential component of executive processes. Ogden et al. used three timing tasks: temporal generalization with a 400-ms standard and comparisons ranging from 100 to 700 ms, reproduction of intervals of 400, 600, 800, and 1000 ms, and verbal estimation of durations ranging from 77 to 1183 ms. In addition, participants received psychological tests assessing performance on the four executive functions described above. On temporal generalization, the number of correct responses was significantly correlated with scores on measures of updating and access. Reproduction accuracy (the deviation between the time reproduced and the target time) correlated significantly with switching ability, and variability in reproduction correlated negatively with updating and access.

The authors further split their adult participant group into two subgroups (high and low scores), using the results on tests of memory span, access, and inhibition. Higher scorers on memory span and access had steeper temporal generalization gradients than did lower scorers, but inhibition had no effect. When the gradients were simulated using the MCG model (see Chap. 4 for discussion), the main difference between the high and low memory span and access groups was in *c*, timing/memory variability, which was lower (i.e. more precise) in high scorers than low scorers. For reproduction performance, splitting the groups in the same way did not result in any significant difference, and for verbal estimation, the only effect was an interaction between access and stimulus duration, with participants with higher access scores producing shorter verbal estimates.

The article by Ogden et al. uses the common-sense notion that different timing tasks might well use different cognitive processes, rather than presupposing that the same psychological mechanisms are used regardless of the task employed, something which is consistent with quantitative models like those discussed in Chap. 4. An article in a similar vein is Droit-Volet et al. (2015). As in the previous studies with Zélanti as an author, children (5 and 8 years of age) as well as adults were used as participants. The three tasks used were temporal bisection, temporal generalization, and reproduction, and all tasks involved two duration ranges, one short (from 0.4 to 0.8 s) and another longer (4 to 8 s). The psychological tests assessed five different psychological processes: working memory, attention concentration, selective attention, attention inhibition, and processing speed. As in the other articles of this type from the Droit-Volet laboratory, within-group *z* scores for the psychological tests were entered into correlational analyses of measures of mean performance and variability. Many significant correlations between psychological test scores and timing performance were observed, and Droit-Volet et al. (2015) sought to simplify these using hierarchical regression analysis. With this method, temporal sensitivity in bisection (in terms of the Weber ratio) was negatively related to attention concentration for both the short and long duration ranges, but the same variable was positively related to the location of the temporal generalization peak only for the longer duration range. The spread of the generalization gradient was related to selective attention. In reproduction, the mean time reproduced was related to processing speed in the short condition but working memory in the long condition, and variability was related to processing speed in both the short and the long conditions.

All of what I have called predictive studies have produced complex results, the details of which are difficult to summarize simply, and this type of work is obviously still in its infancy compared to research using approaches such as dual-task methodology. Nevertheless, certain trends are evident. General intelligence, linked to processing speed and perhaps working memory, appears to be implicated in a number of timing tasks, although this may be something timing shares only with other types of discrimination not involving time. Commonly used timing tasks including bisection, generalization, verbal estimation, and reproduction seem to tap into different processes, and sometimes some common ones as well, and judgements of multi-second durations may not be controlled by the same processes as sub-second durations. Intuitively, these results are perhaps not surprising, although saying this does not disparage the ingenuity and effort that has been expended in demonstrating these facts. Perhaps predictive studies offer the key to understanding some of the between-group differences discussed in Chaps. 7 and 8 regarding the performance of children of different ages or between student-age adults and those considerably older. Only further research will tell, but my personal view is that the predictive methodology discussed in this section holds considerable potential.

Emotion

Research published over the last 10 years on the effects of emotion-provoking stimuli on the perception of duration, much of it conducted by Sylvie Droit-Volet and associates, has attracted considerable interest, but in fact, studies of emotion and timing date back much earlier. In an unusual study, which may be closer to a laboratory simulation of the "emergency time" discussed in Chap. 2 than recent research on emotion, Langer, Wapner, and Werner (1961) tested the effect of potential danger on time judgements in a single experimental participant who operated a moving trolley whilst blindfolded. The task was to produce an interval of 5 s, during which the trolley moved continuously. In a "Danger" condition, the trolley moved towards the edge of a stairwell, towards a precipitous drop; in the "No Danger" condition, the movement was away from the stairwell, down a corridor. On average, the 5-s time period was produced as 3.5 s in the "Danger" condition, but as 4.11 s in the "No Danger" case. Thus, the time interval judged to represent 5 s was shortened, presumably as a result of the fear engendered by this apparently dangerous condition.

A more modern study, by Angrilli, Cherubini, Pavese, and Manfredini (1997), used a more conventional technique, with participants judging the duration of stimuli taken from the IAPS (International Affective Picture System; see Lang, Bradley, & Cuthbert, 2008, for a recent version). The IAPS is a large collection of images which have been rated for arousal and valence (the degree of pleasantness or unpleasantness). Participants were required to estimate, using an analogue scale, or to reproduce picture durations of 2, 4, and 6 s. The pictures varied in valence and arousal level, with high and low values of both dimensions used to produce four sets of pictures, although some neutral pictures were also used as "fillers." Both methods produced underestimation of all durations relative to their real values, but the underestimation was greater with the reproduction method. Results from the study were rather complex. When high-arousal pictures were shown, the negative-valence pictures were underestimated to a lesser degree than those that were positive, whereas when low-arousal stimuli were displayed, it was the positive-valence pictures which were less underestimated. The authors interpreted this pattern of results in terms of two mechanisms. One is arousal-driven, and operates in high-arousal situations; the other, operating in low-arousal situations, is attention-driven. Of course, to account for their results, this explanation necessarily supposes that (a) negative high-arousal stimuli are more arousing than positive stimuli, and (b) in low-arousal conditions, the positive stimuli are "less interesting," which results in less processing of content, and thus they are judged as longer.

A study rather similar to that of Angrilli et al. was conducted by Noulhiane, Melia, Samson, Ragot, and Pouthas (2007), this time using sounds. Their stimuli were taken from an auditory equivalent of the IAPS, the International Affective Digitalized Sound System (Bradley & Lang, 1999). In their first experiment, individuals reproduced the duration of auditory stimuli 2, 4, and 6 s long, and stimuli were selected to provide sets which produced positive high-arousal, positive low-arousal, negative high-arousal, and negative low-arousal sounds, as well as some that were neutral. As in Angrilli et al., the combination of valence, arousal, and different durations produced a complex pattern of results. Neutral sounds were reproduced as shorter than both the positive and negative low-arousal sounds at the 2-s and 4-s durations but not at 6 s. Positive stimuli were reproduced as longer than negative stimuli, and high-arousal stimuli were reproduced as shorter than those of low arousal, although both effects were found only at the 2-s duration.

Noulhiane et al. performed a second experiment using verbal estimation of duration, with durations of 2, 3, 4, 5 and 6 s. Results were once again rather complex. Neutral stimuli were judged as shorter than both positive and negative low-arousal stimuli at durations of 2–4 s, but not at 5 or 6 s. Negative sounds were perceived as longer than positive sounds, and low-arousal sounds as longer than those of high arousal, but only at the 2-s duration. The authors concluded that emotional processes affected the judgement of duration only at short intervals, up to 3–4 s.

Lambrechts, Mella, Pouthas, and Noulhiane (2011) conducted a study using reproduction of visual stimuli lasting 2, 4, or 6 s, including IAPS pictures judged to be pleasant, unpleasant, or neutral, and also used a "content-deprived" grey square. At 2 s, all pictures were reproduced as longer than the square, although there was no effect of picture valence. At 4 s, there was no difference between the content-deprived pictures and the others, although neutral pictures were judged as shorter than pleasant and unpleasant pictures, and at 6 s there were no differences between any of the stimuli.

Overall, then, this set of experiments produced results which are difficult to simply summarize, except to say that emotional effects on time perception are more likely to occur at short than at long durations.

Most of the recent interest in emotion and timing, however, has been stimulated not by work such as that of Angrilli et al., but by a report published by Droit-Volet, Brunot, and Niedenthal (2004). The task used was bisection, with a 400/1600-ms *Short/Long* standard pair presented in the form of pink ovals on a computer screen. Next, participants received test stimuli in the form of female faces with angry, happy, sad, or neutral expressions, which they were to judge as more similar to the *Short* or the *Long* standards, as indicated by S and L responses. The logic here is that the use of neutral stimuli (ovals) to represent the standards, and potentially emotion-provoking stimuli in the test phase, introduces a sort of "state change" between training and testing. If, for example, the pacemaker of a hypothesized internal clock runs faster in the presence of angry faces than neutral ones, this difference is compared with the neutral oval condition, and would be manifested as a displacement of the psychophysical function, for example, with a greater number of L responses overall in the angry face condition than the neutral condition, and consequently a lower bisection point.

In accord with this idea, Droit-Volet et al. found that pictures of faces with emotional expressions tended to be judged as longer than pictures with neutral expressions, but the largest effect occurred with angry faces. The authors discussed a number of potential interpretations of these findings, but their preferred explanation was in terms of arousal: the emotions provoked by the emotional faces, particularly those that were angry, increased arousal and thus increased the speed of the pacemaker of a hypothesized internal clock (see Chaps. 2 and 3 for background). However, they admitted that this interpretation was not conclusive, as only a single duration range had been used. Testing of pacemaker speed interpretation requires more than one duration range, and effects of emotion-provoking versus neutral stimuli should be greater at the longer versus the shorter duration range (see appendix to Chap. 3).

Gil, Neidenthal, and Droit-Volet (2007) used a similar procedure with children 3, 5, and 8 years of age. The bisection tasks involved *Short/Long* pairs of 400/1600 or 600/2400 ms, thus providing the potential for testing the arousal interpretation discussed above. Ovals were used as standard durations and angry or neutral female faces as test stimuli. In all age groups and for both duration ranges, the psychophysical functions when the angry faces were presented were shifted to the left relative to those obtained when neutral faces were used. The typical effect of increasing temporal sensitivity with increasing age (discussed further in Chap. 7) was also observed. There was no significant effect of duration range; this may have been because the two duration ranges were close to one another, although some analyses have suggested a greater difference between the perceived durations of angry and neutral faces when stimuli were longer, supporting the authors' interpretation of the results in terms of an arousal (i.e. pacemaker speed) effect.

Effron, Niedenthal, Gil, and Droit-Volet (2006) replicated the Droit-Volet et al. (2004) findings, showing that angry faces tended to be judged as longer than those with a neutral expression, and this was also the case, but to a lesser extent, for happy faces. However, when participants were required to hold a pen in their mouths so that imitation of the facial expression was inhibited, the effect disappeared, indicating that an important factor in determining these effects is whether the emotional states perceived are embodied in the participant. For other work on potential relations between bodily states and time judgements, see Meissner and Wittmann (2011) and Pollatos, Laubrock and Wittmann (2014).

Another qualifying factor regarding the effect of emotional faces on time perception is the specific task used to measure the effect. Gil and Droit-Volet (2011) conducted an elaborate study in which tasks of bisection, temporal generalization, and verbal estimation of duration, production, and reproduction were used to compare the perceived durations of angry and neutral faces. The bisection task produced results similar to those in Droit-Volet et al. (2004), but temporal generalization showed no effect. Verbal estimation of duration generally showed longer duration estimates for angry than for neutral faces, and a "production-by-waiting" task (Penton-Voak, Edwards, Percival, & Wearden, 1996) revealed that productions were shorter when accompanied by angry faces than by neutral faces, a result indicative of a longer perceived duration of angry faces. In contrast, a reproduction task showed no effect of face type. For reproduction, an angry or neutral face served as a target stimulus (with durations ranging from 400 to 1600 ms), with participants terminating an oval stimulus to produce the same perceived duration as the face. The dependence of the emotional effect on the procedure used to measure it is worrying, and has no clear explanation. Gil and Droit-Volet (p. 361) suggest that the "magnitude of temporal distortion generated by the perception of emotional facial expressions was most evident in temporal tasks in which complex cognitive processes did not interfere with the emotional effect." It is unclear, however, why a temporal generalization task should be considered more cognitively complex than one involving bisection. Models of these tasks (discussed in Chap. 4) would tend to suggest either no difference or that the bisection task was cognitively the more complex, and so the dependence of the emotional face effect on the procedure used remains somewhat a mystery.

Although many of the studies from Droit-Volet's laboratory involved pictures of faces, other stimuli were sometimes used. For example, Gil, Rousset, and Droit-Volet (2009) used a bisection procedure, with a 400/1600-ms *Short/Long* standard pair, and with the standards represented as white ovals. Participants then received test stimuli in the form of pictures of foods which had been previously rated as liked or disliked, as well as ovals. Psychophysical functions from both food types were shifted to the right compared with the ovals—that is, the durations of the pictures were judged as shorter—but the disliked food showed a greater shortening effect than food that was liked. The authors interpreted the results in terms of attention. Although the arguments are not spelled out in detail, presumably the idea is that attention is diverted away from time and to the "content" of the pictures, and this effect is more marked when the foods are disliked.

A study by Droit-Volet, Bigand, Ramos, and Bueno (2010) investigated the effects of what was considered to be emotion-provoking music on time perception. The task once again was bisection, this time with *Short/Long* pairs of 0.5/1.7 or 2.0/6.8 s. The training stimuli were auditory stimuli in the form of sine waves, and the test stimuli were these same sine waves or musical excerpts presented in either a major or minor key, with the major/minor distinction intended to represent the dimension of happiness/sadness. Psychophysical functions from both types of music were shifted to the right relative to the sine wave control condition, that is, the musical samples were judged as shorter. As in the case of liked/disliked food, the authors' interpretation of the results employs the notion that attention is directed away from time to the non-temporal content of the musical stimuli, thus shortening their apparent duration.

Droit-Volet, Ramos, Bueno, and Bigand (2013) conducted a more elaborate study of perceived duration of musical stimuli involving variations of tempo (fast versus slow), musical structure (played normally or backwards), and comparison of tonal versus atonal musical sequences. Musical samples played faster were judged as more arousing and appeared to last longer than music at slower tempi. In addition, music judged as pleasant was perceived as lasting for shorter periods than unpleasant music. Some combination of arousal effects (for fast versus slow) plus attention effects (for pleasant versus unpleasant) was proposed to account for the overall pattern of the results.

Droit-Volet, Fayolle, and Gil (2011) also looked at the effects of mood induced by films on temporal bisection. Participants received temporal bisection tasks with *Short/Long* standards of 200/800 or 400/1600 ms, in the form of blue circles presented on a computer screen. The bisection task was carried out twice, once before experiencing a film and once after. The films were intended to evoke fear or sadness or to be neutral. Psychophysical functions shifted to the left when the film provoked fear, suggesting that stimuli were judged as longer after the induction of fear. The sad film had no effect, even though individuals reported an increase in sadness after viewing, and the neutral film also had no effect. In the fear-induced condition, the effect on bisection points was larger in the 400/1600-ms than the 200/800-ms case, suggesting a multiplicative effect based on arousal.

The effect of mood on temporal bisection performance was investigated another way by Gil and Droit-Volet (2009). In this study, participants were divided into three groups based on their Beck Depression Inventory scores (BDI; see Beck, Steer, & Carbin, 1988, for discussion), to provide groups with no depression, mild depression, and moderate depression. Participants were tested on bisection with 400/1600-ms *Short/Long* standards. Psychophysical functions were shifted to the right in the mild- and moderate-depression groups compared with the no-depression group, although the effect was significant only for the moderate- versus no-depression comparison. Analyses of bisection points confirmed this result: mean values for the group with moderate depression were significantly larger than for the no-depression group. However, there was no effect of depression on temporal sensitivity, as assessed by the Weber ratio. These results suggest that the test durations were judged as significantly shorter in the moderate-depression group than in the group without depression. The authors conclude "that this temporal shortening effect in depression would be due to the speed of the internal clock that ran more slowly in our depressive participants" (p. 174). However, this interpretation is problematic from the perspective of internal clock theory. Because individuals were always in one group versus another, even if there were differences in internal clock speed between the groups, the standard durations 400 and 1600 ms would have been encoded using the same clock speed as the test stimuli. It is thus unclear how a difference in internal clock speed between groups, even if present, would have produced the obtained displacement of the psychophysical functions. Wearden and Jones (2013) provide an extensive discussion of the interpretation of between-group differences on timing tasks. In the case of Gil and Droit-Volet's (2009) study, Wearden and Jones' reasoning would suggest that the effect was due to differences in memory of the standards in the different groups (e.g. being remembered as shorter in the moderate-depression group) or the result of differences in decision processes, such as an increased tendency for a Short response in the moderate-depression group.

The studies reviewed above suggest the involvement of both arousal-based and attentional mechanisms in explaining the overall pattern of results obtained when emotion-provoking stimuli are used in time perception tasks. Evidence for an arousal-based effect comes from the judgements of emotion-provoking stimuli as increasingly different from those of neutral stimuli as stimulus durations become longer, a multiplicative effect of pacemaker speed, discussed in the appendix to Chap. 3. In fact, although there is some evidence for such effects in the data discussed previously, it is rather weak, and Gil and Droit-Volet (2012) attempted to obtain more conclusive evidence for arousal effects.

Their rather complex study used IAPS pictures, and varied both arousal level (high or low) and type of emotion provoked (disgust, sadness, or fear). Recall that IAPS pictures have been rated for their arousal-provoking characteristics, so it was possible to choose appropriate stimuli from the IAPS set. Instead of the usual method of bisection, verbal estimation of duration was used, with a particular focus on whether different stimulus types produced estimates which differed in slope (indicating a multiplicative arousal-based pacemaker speed effect; see appendix to Chap. 3), or a constant additive intercept effect which might reflect attention-based switch processes. In addition to the different stimulus types used, the range of durations was varied between 200 and 800 ms or 400 and 1600 ms. The results were complex, and the interested reader should consult the original article for a complete description. Briefly, when effects of stimulus type were obtained, they were generally manifested in terms of differences in slope between the conditions compared, suggestive of an arousal-based multiplicative effect similar to that with a change in pacemaker speed between conditions. However, the content of the pictures also played a role, possibly by way of an attentional mechanism.

Although a number of different emotions have been explored in the work of Droit-Volet and colleagues, a consistent theme is that negative emotions such as those provoked by angry faces are more effective stimuli for producing distortions of time than positive emotions. Droit-Volet and Meck (2007) show data from a range of emotional states in their Fig. 3 (p. 510), where anger, fear, sadness, disgust, happiness, and shame are contrasted with results from studies using both children and adults. The role of negative, usually fear-provoking stimuli on time perception has also been studied in a series of experiments by Tipples (2008). For other work showing the powerful effects of negative stimuli, see Grommet et al. (2011) and Droit-Volet, Mermillod, Concenas-Silva, and Gil (2010).

Tipples (2008) used a method very similar to that of Droit-Volet et al. (2004), with angry, fearful, happy, or neutral expressions. Pictures with angry expressions were consistently judged as lasting longer than the others, and the extent to which this was true at the individual level correlated with the degree of negative emotionality felt by a person viewing the stimuli. However, Tipples (2010) also found what appears to be the opposite effect on bisection performance of viewing taboo words, in which these words were judged as shorter than neutral words. A possibility raised in an article by Kuhn and Tipples (2011) proposes that the different effects of faces displaying different emotional expressions may be related to the way they are viewed. The authors used an eye-tracking method to examine gaze direction in a study where the direction of gaze of a face cued a subsequent stimulus location. Individuals, when searching for a threatening target, were more likely to follow the gaze of a fearful face than a happy one, whereas the difference disappeared when they searched for a pleasant target. While not directly relevant to the facial expression and emotion literature, this result suggests that the use of eye-tracking methodology might offer some insight into why faces with different expressions affect time judgements differently. Droit-Volet and Meck (2007) also discuss the possibility that negative stimuli are particularly effective in producing changes in time judgements, as in real life they may signal threats, and may thus activate areas of the brain related to the processing of fear-provoking stimuli and reactions to them.

Overall, the studies on emotion and time perception suggest that the induction of an emotional state through the presentation of emotion-provoking pictures can influence duration judgements. However, some caution is in order. Firstly, the effects are often very small, suggesting that duration judgements are changed by a few percentage points at most. In addition, the strongest effects generally appear to result from stimuli which are likely to provoke negative emotions, such as fear and anger: Evidence for the effects of other emotions, particularly positive ones, is much patchier. Effects also may not be manifested for stimuli more than a few seconds long. Furthermore, even the effect of negative emotions seems to depend, for reasons which are not completely clear, on the procedure used to test it. The most effective technique seems to be the temporal bisection method used by Droit-Volet et al. (2004), where the stimuli used as the standards are neutral and the test stimuli are potentially emotion-provoking.

Finally, the interpretation of the effects remains problematic. Part of the reason is logical: perceived durations are lengthened by arousal effects, whereas durations appear shorter with attentional effects. To illustrate this point, consider the results in Tipples (2010). He found that presentations of taboo words were judged as shorter than presentations of neutral words, so presumably the differential processing of the "content" of those words was responsible, diverting more attention away from timing for the processing of taboo words than for neutral words. Indeed, this was his explanation of the results. But suppose Tipples had obtained the opposite finding, with the taboo words judged as longer. Now the effect would be explained by the increased arousal provoked by the taboo versus the neutral words, a perfectly plausible interpretation given the effects of provoking negative emotions in other studies. The problem is that since increased attention to content diverts attention away from timing, *shortening* perceived duration, and increased arousal causes the internal clock to tick faster, thus *lengthening* perceived duration, any possible result can be accommodated simply by changing the balance of these two opposing processes. How can this problem be resolved? One way of attacking it might be to have some independent measure of arousal or attention. Given that the IAPS pictures have been classified in terms of the arousal they provoke, these can be used to manipulate arousal, as in Gil and Droit-Volet (2012). However, providing an appropriate independent measure of how much attention is paid to the "content" of a stimulus presented rather than its duration remains a difficult challenge, although techniques like eye-tracking (see Kuhn & Tipples, 2011) may help.

Summary

Attention to time has long been thought to play a central role in determining time judgements, often employing the idea that accumulation of pulses from a pacemaker-like device will occur only if attention is paid to time, with attention needed to maintain the connection between the pacemaker and its accumulator. Many studies have examined the effects of presenting a non-timing task at the same time a timing task is being performed, and these almost always show a shortening effect on time judgements, suggesting that diversion of attention away from timing does produce a "loss of pulses" from some clock-like mechanism. However, the effect seems to be dependent, in ways which are not yet completely clear, on the type of non-timing task used. Related to this is the issue of whether performance of the timing task also interferes with performance on the non-timing task. Other studies use instructions to "divide" attention; similarly, paying less attention to time decreases time judgements.

A smaller number of studies attempt to predict performance on timing tasks using performance on other cognitive tasks. Results are complex, and the predictor variables are not always the same from one study to another; however, working memory, attention, information-processing speed, and aspects of executive function have all been found to be predictive of timing performance.

Many studies have found that emotion-provoking stimuli, particularly those provoking negative emotions, appear to last longer than neutral stimuli, although the effects may be small, dependent on the measurement method used, and present only for short durations. The customary interpretation of such effects is that the emotion-provoking stimuli increase arousal, thus speeding up the pacemaker of an arousal-sensitive internal clock.

6

Retrospective Timing and Passage of Time Judgements

In this chapter I discuss so-called *retrospective timing* (what Ivry & Hazeltine, 1992, neatly label as "timing without a timer") and passage of time judgements—judgements of how fast or slowly time seems to pass in some situations, as opposed to judgements of event duration.

Retrospective Timing

In 1976, Hicks, Miller, and Kinsbourne introduced a distinction that has since become central to our understanding of human timing, that between *prospective* and *retrospective* timing. Prospective timing is the situation in which people are informed in advance that time judgements are a focus of the task in hand. Retrospective timing is the situation in which unexpected questions about time are asked, such as "How much time has passed since you started reading this paragraph?" By definition, it is assumed that all timing by non-human animals is prospective. Hicks et al. trace the distinction back at least to William James (1890), although whether James' use of the term "in retrospect" is the same as that used today is perhaps questionable, as it seems to involve aspects of

J. Wearden, *The Psychology of Time Perception*,
DOI 10.1057/978-1-137-40883-9_6

both retrospective time judgements and passage of time judgements. To quote James

> In general, a time filled with varied and interesting experiences seems short in passing, but long as we look back. On the other hand, a tract of time empty of experiences seems long in passing, but in retrospect short.

In fact, what came to be known as retrospective timing had been studied before 1976, usually under the term "remembered duration," as opposed to "experienced duration" (prospective timing; see Block, 1974, for discussion).

I should note, firstly, that the terms *prospective* and *retrospective* timing are misleading: almost all time judgements made in experiments are retrospective in the normal sense of the word. For example, deciding whether a stimulus has the same duration as a standard in temporal generalization is almost always made after the stimulus has finished (but see Klapproth & Müller, 2008; Klapproth & Wearden, 2011), as are other types of judgements such as the verbal estimation of duration. The term "remembered duration" as a substitute for retrospective timing is also less than ideal, as memory for the duration of stimuli and other events can be the focus of interest even in a study involving prospective timing; see, for example, Wearden and Ferrara (1993) and Ogden, Wearden, and Jones (2008), both mentioned in an earlier chapter. Likewise, the term "experienced duration" for prospective timing carries with it the implication that duration is not experienced at all when individuals are not alerted that it is a feature of the task, which is surely false. However, given that the terms *retrospective* and *prospective* have become standard, it would be confusing to introduce a new terminology, albeit a more accurate one, so I will continue to use these terms as Hicks et al. (1976) proposed.

Almost all contemporary researchers regard prospective and retrospective timing as different in the sense that the mechanisms responsible for them are different, and some differences between retrospective and prospective time judgements are discussed later. Ivry and Hazeltine (1992) summed this up nicely, proposing that prospective timing is "timing with a timer," whereas retrospective timing is "timing without a timer." One might suggest that prospective timing is performed by some specific timing mechanism, like an internal clock of the type that SET proposes; when individuals are informed that timing is a central feature of the task, they presumably "deliberately" activate whatever timing mechanism they have. In contrast, when retrospective judgements are unexpectedly called for, it is likely that they have not carried out any deliberate timing, and so the duration judgements must come from something else. But what exactly is this "something else"?

In a short but influential volume, Ornstein (1969) provided a possible answer to this question. with his "storage size" metaphor. Ornstein was critical of early attempts to develop clock-based models of timing (such as Treisman, 1963), and instead proposed that all timing, at least in humans with whom he was exclusively concerned, was based on more general cognitive processes, and involved what he called "storage size." In his own words (p. 41), "we will try to relate the experience of duration of a given interval to the size of the storage space for that interval in general information-processing terms. In the storage of a given interval, either increasing the number of stored events or the complexity of those events will increase the size of storage, and as storage size increases the experience of duration lengthens."

In fact, the idea that retrospective time judgements are based on properties of remembered events was not new, and Ornstein acknowledges Guyau as having proposed a similar idea in the nineteenth century. As Guyau (1890) writes

> The apparent length of time appreciated at a distance grows as a function of the contrasted and intense differences perceived in the events recalled (p. 91, my translation).

According to Ornstein, the size of storage is not determined merely by the amount stored, but also by the manner in which it is stored, and in a series of nine experiments he attempted to validate the storage size idea and to explore its limits. Although Ornstein's storage size metaphor is well known, the actual experimental work he reported in his book is very rarely described, so I will devote some space to it here.

All Ornstein's experiments involved what we would now call retrospective timing, as participants were unaware that time judgements would be required. In addition, all except the first involved some sort of *relative* judgement, in that the duration of one condition was always contrasted with that of another, usually by means of an analogue scale. Therefore, if the standard duration was represented by a line of some length, the participant was required to indicate the duration of another condition by marking another line to make it longer or shorter than, or the same as, the standard.

In Experiment I, Ornstein conducted a study similar to that of Frankenhauser (1959). People listened to 9 min and 20 s of repeated 500-Hz tones, presented at different rates, 40, 80, or 120 per minute. The tones were presented at a regular or irregular pace, between conditions. The regularity of the tones had no effect on duration estimates, but these were systematically affected by tone rate, being shortest at 40/min, longest at 120/min, with the 80/min rate between the two, although all estimates generally underestimated the actual duration.

Experiment II used polygons of differing complexity, with 4 to 36 interior angles, and these were rated in complexity from 1 to 18. The stimulus closest to a complexity of 9.5 was used as a standard, and this was presented in contrast to one of the others (sometimes one less complex, sometimes one more complex), with both presented for 30 s. A line representing the duration of the standard was presented, and the participant then marked the perceived duration of the comparison stimulus on another line. Subjective duration increased with increasing stimulus complexity, but plateaued at stimuli with around 12 interior angles, with no further increase in duration as the number of angles continued to increase.

The material for Experiment III consisted of two audio tapes, each involving repetitions of ten different sounds (a zipper or tearing paper, for example), and the total duration was 5 min. The tapes differed only in the arrangement of the sounds. In one ("easily codable"), the same stimulus was presented 20 times in succession, followed by the next sound, and so on. In the other ("random"), the stimuli were presented in random order. Participants received both tapes in a counterbalanced order across the group. The random tape was judged to last 33 % longer on average than the "easily codable" one.

Experiment IV repeated the procedure of Experiment III, but with complex visual stimuli (a van Gogh painting, for example), and a similar result was obtained, in this case a 36.2 % increase in subjective duration for the random stimulus sequence. The four experiments taken together show effects of the quantity or complexity of presented material on duration judgements (Experiments I and II), and the effects of different organization of the same material (Experiments III and IV).

Experiment V involved participants learning a pursuit rotor task, with different degrees of practice, or a control motor task (mirror drawing). The idea here was that the more practised individuals would judge a constant 2-min period of the pursuit rotor as shorter than the other two groups, as experience with it reduced the amount of cognitive activity the task required. The standard condition was a 2-min presentation of the "random" tape used in Experiment III. The results confirmed the hypothesis: the duration was overestimated by 9 % in the practised condition, compared with 56 % and 42 % in the short-practice and no-practice conditions.

In Experiment VI, participants learned to label segments of a 1-min 40-s film of a dancer performing 26 dance movements. Three films were produced, one in which the total sequence was segmented into two sections, another in which it was segmented into six sections, and another in which it was segmented into 11 sections. A 1-min 40-s sequence of the "random" tape from Experiment III served as the standard. Duration estimates systematically increased with increased complexity of segment classification, being nearly 60 % longer in the 11-segment than the 2-segment condition.

Experiment VII involved manipulating the retention of information learned during some event. A pilot study had shown that pairing words with "harsh" sounds caused them to be forgotten more quickly than when paired with "neutral" sounds. A 5-min musical excerpt served as the standard stimulus. Four groups of participants were employed. Two learned a word list paired with harsh words, the other two with neutral words. One group of each type judged the duration of the learning period immediately after it was finished; the other two groups were recalled 2 weeks later, and made their estimates then. Estimates were shorter after 2 weeks than when given immediately but, more critically, were shorter in the harsh condition than the neutral condition after 2 weeks, in accord with the number of words remembered, which was fewer in the harsh group.

The rather complex, but particularly ingenious, Experiment VIII involved a common test stimulus in the form of a complex-looking line drawing which was actually the word "man" written in a cursive script, with the same word also presented at the bottom of the figure in the form of an up-down mirror image. This meant that, normally, the words written in the standard stimulus presentation were not perceived. This test stimulus was viewed for 1 min. A 1-min segment of the random tape used in earlier experiments served as a standard duration. Three groups viewed the test stimulus, but they differed in other ways. Members of two groups saw a 20-s presentation of a complex stimulus consisting of a random-looking set of curved lines. Another group received a "coded set" stimulus in which the upright and inverted script of "man" was separated so that the words could be read. Here the idea was to influence the coding of the test stimulus for this group. Of the other two groups, one received the coded set *after* the test stimulus had been viewed, which meant that the group could change their encoding of the test stimulus retrospectively, using the knowledge that it consisted of the word "man" upright and inverted. The duration estimates were shorter for the groups who received the coded set. Interestingly, however, receiving the coded set shortened duration judgements not only when it was presented *before* the test stimulus, but also *afterwards*, suggesting that the coding of a complex stimulus could be modified retrospectively, and that this change would alter duration judgements.

In the rather similar Experiment IX, the perceived duration of an ambiguous figure was shortened by instructions to code it as an "insect" or a "man," when these instructions were given either before or after presentation of the figure.

Taken together, Ornstein's experiments provide strong evidence for the role of information processing in determining how duration length is judged in retrospective timing. Not only was changing the quantity of information presented found to be important, but the way in which the information was processed was also critical, and this could be altered retrospectively, presumably thus changing the amount of "storage" needed. Ornstein (1969), however, was not completely uncritical of his own position, and drew attention (p. 71) to one of its particular problems, the inability to derive quantitative predictions. One might feel reasonably confident in saying that the storage size in condition X was greater than in condition Y, but it is not possible to say how much greater. In addition, the very idea of "storage size" seems vague, at least in quantitative terms, and Ornstein's own experiments seemed to suggest that the "storage size," whatever it was, could be influenced by a number of factors, making it difficult to predict experimental outcomes in certain cases.

A series of experiments reported in the 1970s by Richard Block, a leading researcher on retrospective timing, although in some ways supportive of Ornstein's storage size hypothesis, suggested limitations and certain necessary modifications. For example, Block (1974), in his Experiment 2, presented individuals with lists of words from different semantic categories. These were presented either in random order, with the different categories interspersed, or in blocks, with all items from one category occurring in sequence, followed by words from another category, and so on. Duration judgements were significantly longer in the blocked condition, even though, presumably, the "coding" of the material was made easier by the blocking of items. As Block notes (p. 158), this suggests that in addition to the "complexity" of the material presented, the degree of organization of items in memory plays a significant role in determining retrospective duration judgements.

Block (1978) conducted another study involving the presentation of visual patterns of differing complexity. In his Experiment 1 he found no effect on duration judgement of the complexity of individual items in the sequence presented, although Experiment 2 found that duration judgements were affected by the complexity of the sequence overall.

The study by Block and Reid (1978) challenged the storage size hypothesis more directly. In the first experiment, participants received two sequences of words: in one they had to count the number of words written in lowercase, and in the other the number of words that were parts of the body. This is a "levels-of-processing" manipulation (Craik & Lockhart, 1972), with the upper/lowercase decision involving "shallow" processing, and the parts-of-the-body semantic processing, considered "deep" processing. The normal result is that memory is superior in the deep versus shallow processing condition, and this was indeed found by

Block and Reed. However, despite better memory in the deep condition, which presumably implies greater "storage," there was no significant difference in duration judgements between the two conditions (a result replicated by McClain, 1983). Block and Reid (1978) suggested that a potentially operative variable was not storage size per se, but rather the amount of "contextual change," the number of *different* types of processing experienced during an interval, echoing the quote from Guyau (1890) above suggesting that remembered *differences* were critical in determining the perceived length of a time period. Block and Reid (1978) tested the possible role of contextual change in a second experiment, which again involved two word sequences. For one (unmixed), the processing was either always shallow or always deep. For the other (mixed), alternate words were processed as either shallow or deep. The authors proposed that the mixed condition involved a greater degree of contextual change, as the type of processing required varied continuously, and in fact they found not only better memory in the mixed versus unmixed condition, but—more relevant for our purposes here—longer duration as well.

Other research like that of Block has suggested that the "storage size" metaphor might need qualification. In a first experiment, Vroon (1970) showed that presenting a greater number of signals in a given time period lengthened retrospective duration, supporting Ornstein's position, and replicating the results of Ornstein's Experiment I. Vroon's Experiment 2, on the other hand, showed that when participants had to respond to the signals given (making a binary choice depending on their frequency, 200 or 2000 Hz), a greater quantity of information transmission then led to *shorter* duration judgements. Other experiments replicated this result, and revealed that this was not dependent on participants making button presses per se: the critical feature was the fact that the stimuli required some active response.

This was further studied by Predebon (1996) in rather complex research, of which I will discuss only a part here. Participants received stimulus sequences which varied in the number of stimuli presented, as well as whether the stimuli was required to be actively processed or just observed, and the duration of the sequence was assessed in either a prospective or retrospective paradigm. Retrospective estimates increased with an increasing number of stimulus presentations during a 48-s interval, regardless of whether these were processed actively or passively, a result differing from that of Vroon (1970).

Retrospective timing remains under-researched relative to prospective timing. In 1997, Block and Zakay conducted a meta-analysis of potential differences between time judgements based on prospective and retrospective timing. In surveying 9000 references to time perception, they found just 20 studies of retrospective timing to include in their review. Admittedly, they used certain exclusion criteria in addition to the requirement that the timing be retrospective, so the number of studies that had been conducted was actually greater than 20. Nevertheless, even when only work with humans is considered, their survey suggests that retrospective timing studies represent just a small fraction of the articles published on time perception. Part of the reason for the relative neglect of retrospective timing research must surely be methodological. In any given experiment, participants can be asked only once for retrospective time judgements, as after that they will have been alerted that time is important, and subsequent judgements are thus likely to be prospective. In the early research discussed above, as a measure for avoiding this problem, participants were asked to make only one judgement, usually of the relative duration of two intervals. With this approach, however, very few data points—usually from judgements of a single interval—are obtained from each participant, rendering the experimental design highly inefficient in terms of data collection, in contrast to work on prospective timing, where a person can be asked to make judgements of a range of intervals (e.g. Wearden, 1995, obtained judgements of up to 24 intervals from each individual tested).

Given the often small effects of manipulations such as changing the content of the material in an interval, or the variability in responses from one person to another, an experimenter risks a large expenditure of experimental time and effort only to find that the differences between the conditions of interest do not reach conventional standards of statistical significance. This situation can be improved somewhat using a method that was employed in several experiments by Boltz (e.g. 1994). Here, participants receive a series of stimuli which are distinct and nameable (e.g. "noise of a car," "baby crying," and "footsteps"), each of which may have a different duration. The typical instruction might be simply to listen to the stimulus sequence, as questions about the stimuli would be asked later. Then, at the end of the sequence, questions about relative or absolute durations are posed (e.g. "how long did the footsteps last?"). This method clearly enables the collection of more data from each person than is possible in experiments like those of Ornstein (1969), but there are still obvious limitations: an individual might reasonably be expected to judge the remembered duration of 5 or 6 distinct stimuli, but not 50 or 60.

Despite the difficulties in carrying out retrospective timing research, a body of work has gradually developed since Ornstein (1969) and the early work of Block and others in the 1970s, and this has not only illuminated the processes involved in retrospective timing, but has also uncovered phenomena not obviously explained by either storage size or contextual change models.

Boltz (1998) looked at the effects of task predictability on retrospective time judgements. In his first experiment, participants performed four simple and familiar tasks (e.g. word-processing or data entry). Groups were either told the order in which the tasks were to be performed, or were given no prior information. Estimates of duration of all the tasks were longer when no information was given, a result reminiscent of Ornstein's Experiment III discussed earlier. In general, other experiments in the same article showed that unpredictability tended to increase estimates.

Another of Boltz's studies produced results which may be more problematic for earlier accounts of retrospective timing. Boltz (1993) showed that a person's expectancy of the length of a task played a role in determining time judgements. Individuals received 36 trials of a music perception task, then had a short break. They were then told that they were one-third, one-half, or two-thirds of the way through the task, before being given another 36 trials (so one-half was actually correct). Participants who were told they were two-thirds of the way through the task, and for whom the second part lasted longer than expected, strongly tended to judge that the second part was longer than the first, whereas those who were told they were one-third of the way through the task tended to judge the second part as shorter. Individuals who were (correctly) told that they were halfway through the task at the break tended to judge the two parts as equal in length. Given that the two parts of the experiment involved the same task, it seems that neither storage size nor contextual change accounts can, in themselves, deal with this result, which was replicated in principle in a second experiment.

Differences Between Prospective and Retrospective Timing

What evidence is there that prospective and retrospective timing really are different? Perhaps some sort of internal clock, which is "deliberately" used for prospective timing, operates "unconsciously" even during retrospective timing, resulting in the two being psychologically similar even if procedurally different. Most current opinion disagrees with this view, and one line of evidence supporting the notion of a difference between the two comes from manipulations of the content of material during an event which is timed either prospectively or retrospectively. Suppose that during a given period of time, the participant is required to perform some sort of information-processing task, which can vary in difficulty over two values, easy and difficult. In prospective timing, the normal assumption is that the more difficult information-processing task would divert attention from timing (this idea is discussed in more detail in Chap. 5), so time judgements would be shorter in the difficult information-processing condition than the easy one. In contrast, ideas like "storage size" would imply more storage in a difficult condition than an easy one, and given that the information stored is the basis for the retrospective judgement, time judgements should be longer in the difficult condition when retrospective judgements are required. This means that the same manipulation, changing the difficulty of information processing, has the opposite effect on the two types of timing, which logically implies different underlying mechanisms.

Numerous demonstrations have shown that changing the difficulty of the task during some time period changes prospective and retrospective duration judgements in opposite ways. Block, Hancock, and Zakay (2010) provide a detailed review, but one simple example comes from Wearden, O'Donoghue, Ogden, and Montgomery (2014). Participants received displays containing the digits 0 through 9 in a 3 × 3 matrix, with one digit missing, and were required to identify the missing digit. In one case (easy), the digits were displayed in rows in numerical order, so finding the missing digit was easy. In another, the digits were randomized in the display, so the search task was considerably more difficult. Numerous displays were presented, and participants were asked how long the entire sequence lasted, using either a prospective or retrospective paradigm. Judgements were longer with the more difficult search task when the judgement was retrospective, but shorter when it was prospective (see also Block, 1992).

In another meta-analysis, Block and Zakay (1997) noted other differences between duration judgements collected with prospective and retrospective methods. Retrospective judgements tended to be shorter, and thus less accurate, as most judgements in the work they reviewed were underestimates of real time values. But this does not always occur. Boltz (2005) presented participants with sequences of naturalistic events in either visual or auditory modality. The events had either a coherent, continuous, or incoherent structure (see the original article for details of how these were defined), and were repeated either four or eight times. In Experiment 1, no differences were found in timing accuracy, assessed by a reproduction method, between prospective and retrospective timing when events were coherent or continuous, particularly when eight repetitions were given. A second experiment presented the same events in either the auditory, visual, or audio-visual modality, and once again either four or eight repetitions were given. Reproduction of duration was less accurate in retrospective timing with four repetitions, but the difference disappeared with eight repetitions.

Another difference noted by Block and Zakay (1997) was that retrospective duration judgements tended to be more variable than prospective ones. Furthermore, this difference in variability between the two paradigms increased as the intervals judged lengthened. In addition, the authors analysed the psychophysical slope, that is, the slope of regression between estimates of duration and real duration. This was markedly higher on average for prospective timing (0.74) than for retrospective timing (0.39).

There are a few studies, however, suggesting similarities between prospective and retrospective timing; for example, Brown (1985). In his first experiment, three groups processed a polygon in different ways. One (control) was required only to look at it, another (easy) traced the contours of the polygon, and a third (difficult) traced it while looking in a mirror. Increasing task difficulty shortened estimates in both prospective and retrospective timing conditions. Brown (p. 119) commented that "task demands also distract retrospective subjects from processing temporal information" and (p. 122) "perhaps the most important finding in this research concerns the basic similarity between prospective and retrospective judgement." The implication here is that individuals in a retrospective paradigm *are* processing temporal information, at least to some extent.

As discussed in Chap. 2, a common notion used to account for prospective time judgements is that these are based on an arousal-sensitive timing mechanism, such that temporal accumulation is greater when arousal is high. Boltz (1994) used arousal manipulations to potentially alter retrospective time judgements, suggesting another similarity between prospective and retrospective judgement. Her method involved increasing or decreasing arousal using 5 min of exposure to the sound of repeated loud car horns (arousal increase) or lapping water (arousal decrease). The independent measure of arousal was a person's *internal tempo*. Internal tempo (see Denner, Wapner, & Werner, 1964, for additional discussion) was assessed by asking individuals to tap for 30 s at a pace comfortable for them, with higher tapping rates indicating faster internal tempi. Boltz's arousal manipulation changed internal tempo reliably in both directions, with the car horns increasing it and the lapping water sounds decreasing it, in comparison with a pre-manipulation level. As was often the case in Boltz's experiments, naturalistic sounds were used as stimuli whose duration was required to be judged in a retrospective paradigm, and these sound samples were either coherent, continuous, or incoherent. In Experiment 2, time judgements were assessed by a reproduction method, and coherent and continuous sounds were reproduced as significantly shorter in the group which had its arousal increased compared with the decreased-arousal group and a control condition without an arousal manipulation.

How can performance on retrospective timing tasks be explained? Unfortunately, there are as yet no formal quantitative models of how retrospective timing is performed, but one possible framework is to suppose that retrospective timing depends on the number of "events" or "contextual changes" counted during a time period, an idea very similar to Guyau's notion that perceived "differences" play a critical role in timing, as mentioned earlier. So, for example, the number of events registered during some time period could be compared with a store of previous events and related times (somewhat similar to SET's reference memory; see Chap. 3) to produce the judgement. The idea that we might have some "stock" of the number of events which habitually occur during different time periods has been proposed in a different context (that of passage of time judgements rather than duration judgements) by the sociologist Flaherty, whose work is discussed in the next section. Even supposing that such a store exists, the definition and counting of events and contextual changes remains problematic.

As noted earlier, although studies by Ornstein and others support that idea that increasing the amount of information processing during a time period makes retrospective duration seem longer, this idea does not always work (see Wearden et al., 2014, for examples). In particular, changing the type of information processing from "shallow" to "deep" (e.g. classifying words in terms of their font or their semantic category) does not necessarily increase duration judgements (Block & Reid 1978; Wearden et al., 2014), even though most psychologists would agree that deep processing involves more "mental work" than shallow processing. In addition, the very problem of how events are to be distinguished and counted remains a critical but troublesome issue, in a way rather similar to that of providing a quantitative measure of the "amount of attention" paid, as discussed in Chap. 5. Over a brief period of time, a person may pick up a pen from their desk, or make eye contact with someone and fall in love. Over an hour or two, a person may idly watch a film in which they have no real interest, or watch the birth of one of their children. The events contrasted here were deliberately chosen to illustrate extremes of importance to the observer, but the point of the number of events counted in these cases is clear, and it is hard to imagine that the events contrasted are equivalent.

The difficulty in producing a properly specified model of retrospective timing performance which might enable clear predictions to be made is yet another obstacle to progress in this field, along with the methodological problems discussed earlier. Although unfortunate, it is perhaps understandable that, compared with prospective timing, retrospective timing remains both under-researched and poorly understood.

Passage of Time Judgements

Passage of Time Judgements in the Laboratory

Although the title of Ornstein's 1969 book was "On the experience of time," it was actually concerned with duration judgements in laboratory situations. Here, the "phenomenology" of time, how the passage of time "feels" in some situation, was not discussed. Of course, in everyday life, we often have the experience that the "feel" of time changes from what it normally seems to be, for example, in a traffic jam or a queue. In these situations, there may be no ambiguity about the duration of events—as in a supermarket queue, a clock on the wall may tell us exactly how long we have been waiting—but there may be something especially noteworthy (usually negative) in a 5-min wait in a queue as opposed to exactly the same time spent doing something else. There seems to be no agreed term for the ratings of the "feeling" of time passage in experimental work, but in an article in 2005, I called this a "passage of time judgement."

As pointed out in Wearden et al. (2014), it is a mistake to regard ratings of passage of time as especially subjective, in contrast to the more "objective" measures of duration judgements, and the temptation increases when the duration judgement is assessed by some objective response like a press on a computer keyboard or a verbal judgement in physical units like milliseconds. In fact, passage of time judgements, like duration judgements, are reflections of internal processes, just those of a different sort from duration judgements. All are equally "subjective," although it is possible that they differ along other dimensions, such as reliability or the ease with which they can be manipulated by experimental operations. Wearden et al. (2014) argue that there is nothing radical in this view, and they trace it back to Fechner's celebrated volume "Elements of Psychophysics" (1860).

There are surprisingly few studies of passage of time judgements in the timing literature, even though changes in the experience of time are commonplace in everyday life. I first review various studies looking at passage of time judgements in laboratory situations. One, reported in Wearden (2005), concerned relations between passage of time judgements and retrospective time judgements. If individuals judge that time passes more quickly during event X than Y, how do they judge the relative durations of X and Y retrospectively? If some event is associated with "flying" time, and another with "dragging" time, is the "flying" event judged as shorter, or does the apparent rapidity of the passage of time mean that it contains "more time" than the "dragging" event, and is thus judged as longer, just as a vehicle going at greater speed covers greater distance?

In Wearden (2005), participants either watched 9 min of the film "Armageddon," chosen because it was one of the rare action films virtually devoid of violence, or waited for the same period for the experiment to begin. At the end of this period, they were asked for ratings of passage of time relative to normal. On average, passage of time during the Armageddon period was judged as faster than normal, and time spent waiting as slower. Participants spent 10 min reading, then were asked for a retrospective time judgement regarding the film or waiting period. There was no significant difference between the mean estimates of the two groups, although the Armageddon period was judged as slightly longer. Therefore, it appears that passage of time judgements could differ between two events, without any difference in their duration when judged retrospectively, suggesting that the two types of judgements are, at least to some extent, independent.

More evidence for this came from Wearden (2008b). Individuals watched clips from a film for three different lengths of time. One group was simply asked to watch the film, the other to note when one of the main characters spoke after another character had spoken. After the three clips were presented, participants were asked for retrospective duration judgements and passage of time judgements. In the group who merely watched the clip, passage of time judgements became progressively slower as the clips lengthened, and this was found in the other group as well, but to a much lesser extent, producing a significant difference in the passage of time judgements between groups. In contrast, retrospective time estimates were virtually identical between the two groups, although the judgements increased significantly with actual duration.

An experiment from Wearden et al. (2014) discussed earlier involved manipulation of information-processing difficulty and its effects on retrospective time judgements. In that study, passage of time judgements were also made, and now the effect was the same for prospective and retrospective timing, as the higher information-processing load made passage of time judgements faster in both cases. Another experiment investigated the effects of both information-processing load and contextual change on time judgements. Individuals received displays of letters, and one group had to count the number of times "E" appeared. For another group, the task was to count both "E" and "A," a condition which logically must involve more information processing than searching for a single target. In addition, the letter strings were presented either in a consistent font on a constant background (low contextual change) or with changing font and background (high contextual change). The conditions were combined into four independent groups. Retrospective judgements of the task duration were unaffected by information-processing load or contextual change condition, but the higher information-processing load increased passage of time judgements in both contextual conditions.

Together, these experiments suggest that (a) changing the "content" of a time period, particularly the amount of information processing required, will reliably change passage of time judgements, (b) a greater amount of information processing is associated with faster passage of time, and (c) passage of time judgements and retrospective time judgements are independent in the sense that one can change without any change in the other.

Another experiment reported in Wearden et al. (2014) sought to duplicate in the laboratory the everyday phenomenon that a time interval may "feel" longer than a person knows that it was. When asked how long you were in a queue, for example, you may reply that "It was only five minutes but it felt a lot longer." Wearden et al. (2014) presented individuals with the same film clip four times. After each presentation, they were asked for a duration judgement, a "feel" judgement ("how long did it feel?"), a passage of time judgement, ratings of attention paid, and some hedonic ratings such as how much people liked the clip. Unsurprisingly, duration judgements scarcely changed over the four presentations, but "feel" judgements increased. Passage of time became progressively slower, and was paralleled by decreases in the amount of attention paid and changes in hedonic ratings. Slower passage of time was associated with greater boredom, less liking, more annoyance with the clip, and a decline in engagement with the material in it. It seems, therefore, that changes in the hedonic tone of an event are associated with changes in passage of time judgements, and that "feel judgements," although expressed in conventional time units such as seconds or minutes, are really disguised passage of time judgements and not normal duration estimates.

Passage of Time Judgements in Everyday Life

Several authors, working mostly from the perspective of sociology or occupational psychology, have been concerned with "temporality," that is, feelings about time in everyday situations rather than laboratories. For example, Flaherty (1991) was interested in the difference experienced between what he called "synchronicity"—when experienced time seems to synchronize with clock-measured time—and "protracted duration," when experienced time seems longer than clock-measured time. To investigate this phenomenon, he first analysed 326 narratives from the press or popular books, usually biographies. In addition, he conducted interviews in which 316 undergraduates were asked about their temporal experiences. Based on analysis of this material, Flaherty proposed a "U-shaped" relation between what he called "stimulus complexity" and "protracted duration." When stimulus complexity was low, time seemed to drag, whereas when it was moderate, a state of synchronicity was reached, and experienced time was judged to coincide with clock-measured time. However, as stimulus complexity increased even further, experienced time seemed again to slow, and a state of protracted duration was once again reached.

Flaherty (1993) provides what is perhaps the clearest exposition of his position. He discusses not only protracted duration and synchronicity, but "temporal compression," where time seems to have passed more quickly than clock- measured time. He was particularly concerned with identifying variables potentially related to experiences which change temporality, and how aspects of temporality are learned. Specifically, synchronicity, which is regarded as a normal background state, is "a taken-for-granted and nearly unconscious aspect of temporal experience....[It] is, nevertheless a skill acquired..." (p. 399). Flaherty suggests that this is based on socialization, as well as "clocks, schedules, seasons, and other socially defined regularities" (p. 399). The "normality" of synchronicity results from other things being "normal" or "moderate." For example, synchronicity occurs when (a) the situation encountered is unproblematic, (b) emotional concern is moderate, (c) "cognitive involvement with self and situation is moderate" (p. 400), (d) there is a normal amount of stimulus complexity, and (e) there is a "typical and consequently familiar density of experience per standard temporal unit."

Protracted duration, in contrast, occurs when (a) circumstances are extremely eventful or uneventful, (b) emotions are highly engaged, (c) there is high cognitive involvement, (d) there is intensified stimulus complexity, and (e) as a result, there is "greater density of experience per standard temporal unit" (p. 401). In sum, protracted duration is "generated by the individual's response to subjective involvement with unusual circumstances" (p. 401).

Flaherty also proposed an S-shaped relation between the amount of conscious information processing and time experience. When conscious information processing is low, temporal compression occurs. Increases produce synchronicity, and high information-processing rates then produce the sense of protracted duration (see Flaherty, 1993; Fig. 2, p. 403).

Flaherty was at pains to distinguish different sorts of activities. He suggested that one influence on temporality was whether a "busy interval" is filled with "routine complexity" or "problematic complexity." The latter engages attentional resources, increases the "density of conscious information processing," and therefore causes protracted duration experience. Routine complexity requires less than normal information-processing rates, and thus will result in temporal compression. Thus the type of "busyness" plays a determining factor in temporality, according to his account.

Flaherty and Meer (1994) studied the experience of temporal compression empirically. They were particularly interested in the idea that temporal compression was a feature of memory, so time periods were considered to appear shorter when recollected, with more compression occurring as time since the events increased. The authors conducted a questionnaire study with three different groups: traditional students (mean age 19.8 years), non-traditional students (mean age 38.4 years), and older people (mean age 71.2 years). All were asked to rate passage of time yesterday, last month, and last year, on a five-point scale, where 1 = very slowly and 5 = very quickly. All groups systematically increased their rating of speed of passage of time between yesterday and last year, and there was a small effect of age, with the non-traditional students rating passage of time as more rapid than those in the older and younger groups (effects of ageing on ratings of passage of time are discussed in Chap. 8).

In another article, Flaherty (2003) reported results from interviews with students about how individuals attempted to manipulate their temporal experience. He discussed ways in which they tried to control their experience of the duration, frequency, and sequence of events. He also discussed the ways in which people tried to arrange the time at which events in their lives occurred, and how they managed their time. The results were given in the form of representative quotes, which make them hard to simply summarize, and I will concentrate here only on attempts to manipulate duration. Individuals were motivated to try to "speed up" boring or unpleasant events, and appeared to do this by trying to fill the interval with activities, presumably to distract themselves from the oppressive slowness of time's passage. If the interval was short, these were often rather mundane activities like doodling, writing, or talking. In contrast, people often wanted to make pleasant events last longer, and used techniques such as relaxed breathing or contemplation of the events while they were in progress.

In an unusual study, Flaherty, Freidin, and Sautu (2005) interviewed Argentine students about passage of time during the 5 months after the resignation of President Fernando de la Rúa in 2001. This was a period of extraordinary political upheaval in the country: there were five different presidents within a period of less than 2 weeks, gross domestic product shrank alarmingly, and inflation increased from 0.7 % to 41 %, along with a near doubling in unemployment and the proportion of the population living below the poverty line. About 200 respondents were asked whether time passed quickly, slowly, or synchronically (i.e. in accordance with clock time), or whether they were unable to specify. They were then asked to give reasons for their response. Fifty percent said that the 5-month period has passed quickly, 25.8 % slowly, 19.2 % were unable to specify, and only 3.5 % said that time had passed normally. Those who responded that time had passed quickly often referred to the extraordinary pace of events, for example, "because a lot of things happened in very little time" (p. 404). In contrast, those who thought that time passed slowly tended to focus on the problems of daily life, unemployment, and insecurity, as in "it seems that the suffering is never going to end" (p. 405).

Larson (2004) expanded some of Flaherty's ideas to develop a more comprehensive theoretical framework for the study of temporality. She included his three categories—protracted duration, synchronicity, and compressed duration—but added others. One was "flow," where a sense of timelessness was engendered. Another was "interstitial time," the time that passes (usually slowly and with some dissatisfaction or discomfort) between events; for example, time passing when waiting. Finally, there was "temporal rupture," where as a result of "catastrophe or events that restructure life, time is released from the typical routines of everyday life" (p. 27). One of Larson's foci of interest was in how occupations, and the activities these entailed, changed the experience of temporality. Her framework, the Dynamic Occupation in Time (DOiT) model, was intended to provide a comprehensive framework for the study of temporal experience across a range of situations. The whole picture proposed by DOiT is too complex for full exposition here, but can be illustrated with a few examples. Larson generally follows Flaherty with respect to protracted duration, synchronicity, and temporal compression, although her accounts are more detailed, so I will concentrate on her added categories of temporal experience here.

The sense of timelessness engendered by flow occurs when individuals engage in intensely involving experiences which allow them to focus on the task (rather than the self) and which involve rich but slightly challenging tasks, well matched to a person's capabilities. In contrast, interstitial (waiting or interruption) experiences involve uncertainty and expectation, engrossment in the passage of time, where the focus is directed internally away from stimuli, with an intense, often uncomfortable density of experience, which causes "expanded" time. Temporal rupture, occasioned by life-changing events, involves intense emotional arousal and involvement with the self, as well as a feeling of overwhelming complexity, leading to a distorted "out of time" experience.

Larson's framework is sufficiently broad to cover a range of temporal experiences, including abnormal ones. Although at first it might seem no more than an elaborate classificatory scheme for temporal experiences, it has in fact been used in empirical research. For example, Larson and von Eye (2006) used an experience-sampling methodology (ESM) to investigate temporal experience during the daily lives of 35 students. The students were paged at ten random times each day, and after the pager alert were required to complete a brief questionnaire. Of particular interest, following from the DOiT model, were relations between passage of time relative to clock time (e.g. protracted duration, synchronicity, and temporal compression) and measures of involvement with the novelty of the task they were engaged in at the moment of the alert, their emotional and intellectual engagement with it, their intensity of focus on the activity, and focus on themselves, as well as complexity of the activity and its reliance on skill or innovative actions. Structural equation modelling revealed that engagement and time passage were strongly and mutually linked, each influencing the other. Engagement was predicted by a composite variable comprising novelty, complexity, and skill use during the activity engaged in. Temporality, in fact, was influenced only via engagement, which itself was influenced by the composite novelty/complexity factor. This very sophisticated study is a model for research into temporality in everyday life, and represents perhaps the most technically developed research of its kind at the time of writing.

In a study of everyday life experiences with a different focus, Wearden et al. (2014) conducted a questionnaire study of passage of time in a group of more than 200 students. Questions were asked about the passage of time after the consumption of different drugs, when engaged in various activities, and in different emotional states. Alcohol consumption was associated with fast passage of time, and was the only substance to have a significant effect. Being with friends or family, or partying produced very fast time passage, and being busy or happy sped up the passage of time, whereas being bored, sad, or tired slowed it down. We should note, however, that all these are *reported* effects: for example, the effect of alcohol on timing tasks when administered in the laboratory may not be the same as that reported by the participants in the questionnaire study (see Ogden, et al. 2011, for example).

Wearden et al. (2014) also asked for "time anecdotes," brief reports of situations where time seemed to pass especially quickly or slowly. "Fast time" anecdotes invariably involved two components. One was an external time marker ("I looked at my watch and found it was 3:00 in the morning"), and the other was an inference that time had passed quickly ("so time must have passed really quickly"). People did not *experience* rapid time passage, but instead *inferred* that time had passed quickly, probably because they had paid no attention to it. In contrast, "slow time" anecdotes, which usually involved waiting, reported changes in time experience when individuals were actually in the situation. As such, "slow" and "fast" time reports did not seem to be opposites: slow time reports were accounts of experiences, whereas fast time reports were based on inferences.

An obvious problem with a questionnaire study of this sort is that people may be responding based on conventional ideas or inferences made after the event rather than their direct experience, in contrast to laboratory studies of passage of time judgements, which presumably reflect a person's reaction to the rather arbitrary events presented to them. One way of obtaining what might be a more realistic measure of the passage of time in everyday life is with the use of ESM, as in a study by Droit-Volet and Wearden (2015), who used this method to examine passage of time in two participant samples, one comprising students and the other individuals with an average age of over 70 years. Eight alerts were provided each day, and the questionnaire posed questions about passage of time at the moment of the alert, as well as others about affective state (degree of happiness/sadness), arousal (high/low), the degree of difficulty of the task they were engaged in at the time of the alert, and the extent to which this engaged their attention. In the student group, higher degrees of positive affect were associated with faster passage of time, whereas there was no significant effect in the older group. In contrast, higher ratings of negative affect (i.e. being sadder) were associated with slower local time passage for both groups. Arousal had similar effects in both groups, increasing rated passage of time when higher and decreasing it when lower.

In conclusion, studies of feelings of passage of time in everyday life have suggested multiple factors affecting temporal experience, or reports of temporal experience. Some studies, like those of Larson and von Eye (2006) and Wearden et al. (2014), have been careful to distinguish judgements of temporality or passage of time from judgements of duration, whereas early work by Flaherty did not always make this distinction as clearly. Wearden et al. (2014) questioned whether reports of "fast time," or temporal compression in Flaherty's terminology, were based on experiences rather than inferences, although the consensus seems to be that "protracted duration" can be felt during the event described, so is based on experience, as anecdotes reported by Flaherty and Wearden et al. suggest. Compared with the technical complexity of measures of duration judgements, particularly in the field of prospective timing, attempts to measure passage of time in real life appear rather crude, although data are sometimes analysed using complex statistics. One obvious issue that arises concerns which questions should be asked in order to capture fluctuations in temporality in everyday life and its relations to other variables, and considerable confusion still exists as to what people mean when they talk about time going rapidly or slowly. In particular, there is confusion about what it means to say that "time flies" or "time drags" when the supporting information for this assertion comes from experiments on duration judgements, as I show in Wearden (2015). Flaherty's work suggests, in accord with common sense, that good correspondence between experienced temporality and clock time is "normal" and provides a background against which "abnormal" temporal experiences can be assessed (see also the work on "emergency time," reviewed by Arstila, and discussed in Chap. 3). Indeed, it may be that people spontaneously report fast or slow time (or the other abnormalities noted by Larson) only in unusual circumstances. They can, however, be asked about passage of time in mundane laboratory tasks, as Wearden et al. (2014) show, and their passage of time judgements in these cases can vary depending on task demands. In laboratory settings, as discussed earlier, increasing information-processing demands results in faster passage of time judgements, possibly by diverting attention away from the perception of the passage of time (similar to the diversion of attention away from duration judgements, discussed in Chap. 5), whereas the work by Flaherty, Larson, and colleagues seems to suggest that time is "protracted" (i.e. slowed) by increasing information density in real-life situations. The question of whether passage of time judgements in the laboratory and in real-life situations tap into the same processes, as well as other (mostly unresolved) questions about passage of time judgements, is treated more fully elsewhere (Wearden, 2015).

Summary

Retrospective timing involves time judgements made when individuals are not alerted in advance that questions about time are going to be asked. Ornstein (1969) proposed that retrospective time judgements were based on the "storage size" of events during the time period—for example, the amount of information processing or the number of different types of information processing. Later research raised questions about how "storage size" should be defined, a problem noted by Ornstein himself. Retrospective time judgements tend to be less accurate and more variable than those obtained from prospective timing studies (i.e. those where people are notified in advance that time is an important dimension of the task), although a few studies find little difference.

Passage of time judgements refer to statements about the "feeling" of time, for example, whether it seems to be progressing quickly or slowly, compared with normal. Among the factors that affect reported passage of time in different situations are the amount of information processed, "engagement" with the task in hand, emotional state, consumption of alcohol, and being excited or bored. Passage of time judgements can change when duration judgements based on the same event do not, suggesting that the two types of judgement are not the same. Anecdotes reporting "fast time" generally seem to be inferences rather than experiences (i.e. people do not really feel that time is going fast, but make this statement after the event), whereas "slow time" reports appear to be based on an experience of slowness during the event.

7

Time Perception in Children

A number of issues have been studied in research on time perception conducted with children, various questions have been the focus of diverse investigations, and different theoretical approaches have been used, along with differences in the methodology employed. Like many areas of the psychology of time, studies with children do not represent any sort of unified area of research, and results and theoretical accounts from one area may be of limited usefulness in understanding data from another one.

In this chapter I will distinguish two types of studies. The first is what I will call *Piagetian time psychology*, derived from the work of Piaget, in particular the research contained in the book *The Child's Conception of Time*, initially published in French in 1946, and available in English translation only much later, as Piaget (1969). I will divide Piagetian time psychology into two parts. In the first part I will discuss some findings and ideas from Piaget's own book, and in the second will review some "neo-Piagetian" studies which, although influenced strongly by the work of Piaget, use more modern experimental procedures and different theoretical approaches. Lastly, I will discuss the most recent type of study, heavily influenced by internal clock theory, particularly SET, although not all researchers in this area use SET explicitly.

J. Wearden, *The Psychology of Time Perception*,
DOI 10.1057/978-1-137-40883-9_7

The Child's Conception of Time

Jean Piaget's *The Child's Conception of Time* is a complex and multi-faceted work, unfortunately often difficult for the modern reader to absorb. Not only are we faced with many pages expressing results in a daunting logical formalism, which seems unhelpful to a contemporary reader, but the often-complex experiments employed are only sketchily described. For example, some involve an apparatus in which liquid flows from one vessel into another of a different shape. In other cases, children view dolls displaced in space across a table, and in yet others the children themselves may move (by running along with an experimenter, for example), or describe everyday actions, like going to school. No diagrams or photographs of the experimental arrangements are provided in the book, so the reader can struggle to visualize exactly which events occurred and what they might have looked like to the child. Data are provided in the form of transcripts of individual children's responses to various questions about the experimental arrangements presented. The questions might differ from child to child, so issues obviously arise regarding the representativeness of the children's responses and the degree to which they might have been influenced by prompting from the experimenter or the exact wording of the questions asked.

Despite these problems, Piaget's book provides a rich source of data and ideas, some of which, luckily, have been followed on by others in the Piagetian tradition, almost invariably in studies where the methodology is more clearly described than by Piaget himself, and in which the presentation of results and theoretical analyses are less problematic than those in Piaget's book. Here, I will provide a sketch of some of the work in Piaget (1969), concentrating in particular on issues which have been influential in guiding later research. The reader is warned that the material here is only a superficial summary, and that for the full flavour of the research, the book itself must be consulted.

Piaget was interested in how certain conceptions which are facets of the perception of time, such as the notions of succession and duration, develop with experience as a child grows older. Perhaps unsurprisingly for a French-speaking researcher, he follows Guyau (1890) regarding a child's ideas about time as *constructions* based on world experience, rather than on innate a priori knowledge, as Kant proposed (see Chap. 2 for discussion). The general methodology therefore usually involves testing children of different ages on the same or similar problems to observe the unfolding of a developmental sequence. As typical for Piagetian developmental psychology, the sequence is often classified according to stages, where children would normally be expected to pass through one stage before moving on to the next.

One recurring issue in the book is the difficulty young children have in dissociating *temporal* aspects of an experimental situation, such as how long an event lasted, from *non-temporal* features of the same event. A prevailing theme of the work, and verified by later researchers, is that for the younger children, "more equals more time." This means that judgements about time among the younger children will be strongly influenced by features of the experimental situation, such as distance traversed, quantity of liquid accumulated, size and shape of vessels used, or speed. This does not necessarily always lead to incorrect time judgements: at a constant velocity, travelling a greater distance will take longer, for example, so in this situation, confusion between distance and time will produce the correct response to a question about duration. However, sometimes they are misleading: going faster over a constant distance takes *less* time rather than *more,* so focusing (or "centrating" as Piaget sometimes says) on velocity when questions about duration are asked will tend to lead to incorrect responses.

In some experiments, the child was presented with an apparatus in which liquid could flow from an upper flask ("in the shape of an inverted pear," p. 7) into a cylinder below, with the two connected by a tap. One of Piaget's most striking results, revealed with this apparatus but also with other experimental tests, is that young children fail to grasp what he calls "the equality of synchronous intervals": even though the child might agree that different events start and stop together—such as the changing water levels in the upper and lower flasks—they do not necessarily judge the two events to have lasted the same length of time. In other work, children were given drawings of the water levels in the two flasks, then the drawing of the upper and lower flasks were separated, and the child had to arrange the drawings in the correct sequence. The reader will not be surprised to learn that this complex task was almost impossibly difficult for the youngest children.

Other work discusses the development of the concept of duration, sometimes using the water-level method, sometimes questions about other types of situations. Initially, according to Piaget, the child has no real concept of duration per se, as opposed to the idea that the time things takes depends on other factors, such as the absolute change in the water level in the apparatus, or non-temporal factors such as speed. Some of Piaget's observations are striking, and reveal that the conceptual world of a child with respect to time is very different from that of an adult. For example (p. 37)

> How long does it take you to go home from school? *Ten minutes.* And if you were to run, would you be getting home more quickly or more slowly? *More quickly.* So would it take you longer or not? *Longer.* How much? *It would take ten minutes.*

This example is rich in conceptual confusion. The child, correctly, states that running would get him/her home more quickly, but erroneously states that running would take longer, perhaps an example of more (in this case faster) equals more time. The actual clock time, however, remains the same, despite the previous two contradictory answers. The example also shows something common to many other demonstrations in Piaget's book: non-temporal features of a situation, such as distance traversed or velocity, frequently lead younger children to erroneous judgements of duration. Piaget writes (p. 47) "The belief that time is directly proportional to velocity is rooted in the mistaken idea that activity and the work done can be combined into a single concept." In particular, Chap. 3 of the book, which discusses experiments mostly involving time and movement (sometimes the movement of the child, sometimes the movement of toys across a table), provides a number of demonstrations of the ways in which children confuse distance moved, or speed, with duration. Piaget summarizes this (p. 88) in terms of three rules which can lead to erroneous judgements: "(1) If you go more quickly you necessarily cover more space (i.e., velocity is proportional to distance); (2) if you cover more space, you need more time to do it (hence displacement is proportional to time); and (3) if you go more quickly you need more time because you cover more space. Each of these 'mores' entails the other two."

As they develop, children gradually overcome confusion between velocity, speed, and time, but the progression towards adult "scientific" concepts of time is slow and uncertain. For example, when a child and an adult run through the laboratory, the child might agree that they started and stopped together, but conclude that they did not run for the same length of time because they did not run at the same speed (p. 46). This is another example of the problem discussed earlier, that events which start and stop together are not always regarded by young children as having lasted the same length of time. Piaget attributed this confusion in part to his assertion that distance and velocity are more salient and more easily extracted from the experimental setting than is time: for example, (p. 40) "primitive concepts…are distance and velocity and….it is time which is gradually derived from them…." This position links to earlier philosophical ideas about time from Kant and Guyau discussed in Chap. 2.

Another important aspect of the development of time perception for Piaget is mastery of what he calls "colligation" of temporal intervals, the fact that shorter intervals are necessarily contained within longer ones. For example, in the water flow case, if the water level changes from one level to another, and then from that level to yet another, the first time interval is necessarily contained in, and therefore shorter than, the second: that is, the part is necessarily shorter than the whole. Piaget's Chap. 3 illustrates water level problems showing that children 7 and 8 years of age have great difficulty understanding this colligation of intervals, although from the age of 9, they have much more success with this sort of problem.

Chapter 4 provides an exploration of the development of the child's idea of simultaneity. Initially (ages 4 and 5), the children confuse duration and distance, so for example, when the child and experimenter start and stop running at the same time, the fact that the child is behind the experimenter leads to the conclusion that the child stopped first.

Piaget also used variants of the water flow apparatus, where a large vessel "is allowed to empty…yielding two identical jets of water. The water is collected in small bottles or glasses of different shapes and dimensions" (p. 121). The flow is controlled by a single tap, so the two streams can be observed to start and stop simultaneously. The fact that the water does not necessarily rise to the same level in the vessels which receive it, depending on their shape, causes serious problems for younger children when asked about time. For example, if one vessel is full when the water flow stops and the other is not, children are reluctant to judge that the flow lasted the same length of time, even when they agree that the flow started and stopped at the same time.

In a later chapter, Piaget discusses the child's concept of age. For young children, this is often confused simply with size, or growth. For example (p. 203), a child of four and a half years is asked about the age of members of her family. First she is asked about her younger sister.

> Is she a baby? *No, she can walk.* Who is the older of you two? *Me.* Why? *Because I'm the bigger one.*

The confusion of growth/size leads to a correct response here, but then....

> Is your mother older than you? *Yes.* Is your Granny older than your mother? *No.* Are they the same age? *I think so.* Isn't she older than your mother? *Oh no.* Does your Granny grow older every year? *She stays the same.* And your mother? *She stays the same as well.* And you? *No, I get older.* And your little sister? *Yes!*

Piaget's book provides many striking examples of the fact that a young child's concept of time is very different from that of an adult. In many cases, however, most readers will probably feel that part of the children's problems arise because of the complexity of the experimental situation used, or the complexity of the language in which questions are posed. Fortunately, many of the basic themes of Piaget's book, such as problems distinguishing distance traversed, velocity, and time, have been explored by later researchers, so confirmation of many of his conclusions has been provided by less complex and ambiguous experimental methods, as discussed in the next section.

Neo-Piagetian Studies

Piaget's own studies of time perception in children, although original and innovative, leave much to be desired by modern standards. Fortunately, other researchers followed on some of his ideas using methods much more like those of modern experimental and developmental psychology. Friedman's (1982) edited volume contains some fine examples of neo-Piagetian research, which follows on ideas from Piaget's work using methods which are better described and with clearer theoretical models applied to data.

Several of the studies in the book are concerned with the traditional Piagetian issue of children's problems in distinguishing the duration of events from other aspects of the experimental situation, particularly distinguishing time from space. Richards (1982) first discusses earlier work by Levin (1977), who sought to simplify the complex situations Piaget himself had used when studying time/space interactions. She did this by comparing children's judgements of time in three situations. One was "still time": here, unmoving dolls remained asleep for various periods of time, which were the basis of the child's judgement. These were contrasted with "rotational time," where the dolls rotated on a turntable for various periods of time, such that movement was present but there was no displacement in space. Finally, a "linear time" situation was studied where, as in Piaget's own work, dolls moved through space, so displacement in space might be a factor interfering with time judgements. This was indeed the case, with children having particular difficulty with "linear time" problems because of the interfering effect of spatial displacement on duration judgement. In her 1977 article, Levin introduced a "facet model," which tried to predict the difficulty of duration judgements based on the number and type of potential interfering factors, and this was largely successful, suggesting that these factors had been correctly identified. Later work by Levin (1979) showed that spatial displacement was not the only factor that could interfere with duration judgements, by demonstrating that the size and intensity of static lights also interfered. As Richards (1982, p. 16) puts it, "more on one scale = more on the other scale," echoing the Piagetian slogan that "more equals more time."

Richards' own work was concerned with discerning the rules that children use when confronted with a "linear time" display where trains moved on parallel tracks. The trains could start or stop at the same or different times, run for the same or different lengths of time, and run at the same or different speeds. The problems presented to the children were designed to ascertain the rules that children of different ages were using. For example, an "end-point" rule would suggest that the train further ahead had run longer, or a greater distance, or faster, depending on the question asked. Richards' study involved 5-, 8-, and 11-year-olds as well as adults. Results from these complex situations were themselves complex, but one finding was that 5-year-olds used end-point rules, 8-year-olds no clear rules at all, 11-year-olds used distance, and adults used time.

Levin has perhaps done more than anyone else to systematize developmental changes involved in tasks concerning time and space judgements. In a later article (Levin, 1992), she provided a complex model describing the developmental sequence by which children move from focussing on a single feature of the experimental situation, such as the start or end point of an event, to coordinating starts and ends in an adult-like way. To use traditional Piagetian terminology, young children may "centrate" on a single feature of the experimental situation (perhaps one irrelevant to the question being posed), before learning to "decentrate" and coordinate multiple aspects of the situation in front of them, such as starts and ends, distance, and speed. Levin (1992) describes how these changes occur as children age.

Although the above-mentioned studies suggest progressive development towards adult-type reasoning about time as children grow older, work by Crépault (e.g. 1979; see also Montangero, 1984, for discussion: both articles are in French) suggests that development continues well into adolescence. The task used in Crépault (1979) is a rather complex one: a strip of paper is pulled over a blade, and at the same time a periodic marker repeatedly falls on the strip, producing successive marks. If the period between marks is constant, but the paper moves faster and faster, the space between the marks on the paper increases progressively as the rate of the paper movement increases. Crépault's basic experimental task typically involved comparisons of two strips with different speeds or different patterns of marks. This experimental arrangement obviously comprises multiple events with the potential to create confusion when questions are asked about the speed of the paper, the marks, or time relations between the marks.

Crépault produced a formalization of the relations between distance, time, and velocity, and discovered that these appeared to change throughout adolescence. Between the ages of 9 and 14 years, children's temporal reasoning always involved a negative relation between time and speed: more time equalled less speed. However, other relations changed within this age span: for example, children shifted from believing that greater distance involved lower speed, to associating it with higher speed. Some of the binary relations between pairs of the triad of time, distance, and speed are considered "connaissances generales" (basic knowledge), and the relations can combine to yield new cognitive "objects," "connaissances inferées" (inferred knowledge). Montangero (1984) gives the following example: greater distance implies less time; less time implies greater speed; therefore, greater distance implies greater speed. The "objects" so constructed can be either compatible with "basic knowledge" or incompatible with it, in which case the reasoning structure is unstable and changes by eliminating the incompatible "inferred knowledge" or substituting it for the previous "basic knowledge."

Crépault's work is little known among English-speaking time researchers and, like that of Piaget, has a complexity of methodology and a formalism that may not appeal to many. Perhaps the most interesting general feature of his research, however, is that "correct" (by this he means correctly scientific, or what he calls "Galilean") reasoning regarding relations between time, distance, and velocity emerges only in adulthood, even though adolescents may avoid most of the errors made by younger children in other Piagetian studies.

Studies in the Piagetian and neo-Piagetian tradition are generally dependent on children's ability to understand the questions asked and to respond using appropriate words to describe temporal characteristics of the events of the experimental situation. A study by Tillman and Barner (2015) specifically examined how children acquired and used words describing duration, what the authors called "the language of time." They studied the acquisition of words for conventional time units, and their first experiment involved 3- to 6-year-old children's comprehension of units of time from one second to a year. For example, did the youngest children know that a second is less than a minute, and so on? The 3-year-olds performed at chance level on this sort of task, that is, they could not reliably discern that one temporal unit was longer or shorter than another. All older children performed above chance, with accuracy improving from 4 to 6 years of age. One factor that might be important in children's understanding is the "ranking" of words for time, the knowledge that the difference between a minute and an hour is less than that between a minute and a year. This was indeed found to be a significant factor in children older than 4.

A second experiment complicated the situation somewhat by accompanying the duration word with a numerical label, and this could be "congruent" with the duration—going in the same direction: 3 h versus 2 min, for example, where the larger number is associated with the larger temporal unit—but it could also be "incongruent"—2 h versus 6 min, for example, where the numerical qualifier and the duration measure proceed in opposite directions. Children aged 4 to 7 years were studied. Performance improved with age, but 4- and 5-year-olds performed more poorly on incongruent trials than congruent trials, an indication of their difficulty in separating the numerical label from the duration word, something clearly reminiscent of elements of the Piagetian work described earlier.

SET-Based Studies of Timing in Children

As mentioned earlier, scalar timing theory provides theoretical models which can be fitted to data to yield psychologically meaningful parameters. This means that when comparing the performance of different groups of people—for example, children of different ages—SET-based analyses can provide clues as to why the performance of the groups differs, supplementing statistical demonstrations that reliable differences do occur. Another advantage of a SET-based approach to the developmental psychology of time is that methods associated with SET, such as temporal generalization and bisection, provide a simple approach which may be practicable even with very young children. SET-based models have been widely applied to data from children of different ages, most notably by Droit-Volet and collaborators, but historical precedence must be attributed to McCormack, Brown, Maylor, Derby, and Green (1999). In fact, McCormack et al. (1999) did not use SET itself, using instead a similar theoretical model, but I include them here, particularly as certain of their ideas have been taken up by SET-based analyses.

The study of McCormack et al. (1999) is unusual in that it used not only groups of children of different ages, but also two groups of elderly participants, all contrasted with a student-age group. I will discuss their work with elderly individuals in a later chapter.

In their first experiment, McCormack et al. used a temporal generalization procedure, with a 500-ms tone as the standard duration, and non-standard comparison stimuli spaced in 125-ms steps around the standard. The groups of interest here were 5-, 8-, and 10-year-olds, whose performance was contrasted with that of young adults (mean age 19.1 years). The temporal generalization gradients obtained from student-age participants were (a) peaked at the standard duration and (b) were skewed to the right, such that stimuli longer than the standard by some amount yielded a greater proportion of YES responses than those shorter by the same amount. This is the same result as that found in Wearden (1992) and other studies. In contrast, gradients from the children, although peaked at the standard, were more symmetrical, or in the case of the 5-year-olds, skewed to the left, so that the 375-ms comparison elicited a larger number of YES responses than the 625-ms comparison. The steepness of the gradients also increased with increasing age.

A second study used a bisection procedure, again with auditory stimuli, with standard *Short* and *Long* durations of 200 and 800 ms, with non-standard comparisons spaced in 100-ms steps between these values. The 5-year-olds produced significantly flatter psychophysical functions than the other groups, who did not differ. Bisection points were above the geometric mean (400 ms) but below the arithmetic mean (500 ms), cf. Wearden (1991b).

McCormack et al. modelled their data using an account rather similar to SET. For the temporal generalization task, it was assumed that the perception of each duration was susceptible to "noise," a sort of timing variability, and there was also a response threshold (rather similar to that used in the MCG model of Wearden, 1992; see Chap. 4 for details), effectively representing the degree of similarity between the comparison and standard duration necessary to produce a YES response. In addition, the authors proposed that the memory of the standard was subject to "distortion," and could be remembered as shorter (distortion < 1) or longer (distortion > 1) than it really was. When this model was fitted to the data, the noise parameter decreased with increasing age, although the 10-year-olds produced values almost identical to those of university students. Thresholds also decreased with age, in this case indicating less conservative performance as the children's age increased. The "distortion" parameter decreased with age, although the main effect was a difference between the 5-year-olds (0.87) and the other groups (0.96, 0.96, and 1, in order of age). Bisection was modelled with a noise parameter alone, and once again the value decreased with increasing age, with the main effect the difference between the 5-year-olds and the other groups.

Two studies, this time using standard SET modelling, followed on these results. Firstly, Droit-Volet and Wearden (2001) used groups of 3-, 5-, and 8-year-olds on a temporal bisection task with two duration ranges involving *Short*/*Long* pairs of 1/4 s and 2/8 s. The children received initial training to ensure that the standard *Short* and *Long* durations were well discriminated. For both duration ranges, the younger groups (aged 3 and 5 years) produced flatter psychophysical functions than the oldest group, and this was reflected in larger Weber ratios. Bisection points were close to the arithmetic means of the standard durations for all but the 2/8-s duration range in the 8-year-olds. When the comparison stimulus durations were divided by the bisection point for the different conditions, approximate superimposition was found.

The data were modelled using the Wearden (1991b) model, described in Chap. 4, and a variant. When the basic model was used, the timing/memory variability parameter decreased systematically (indicating less variable timing or memory) with increasing age in all cases. The bias parameter decreased with increasing age in the 1/4-s condition, but was largest in the oldest children in the 2/8-s condition, consistent with their bisection point being close to the geometric mean. A variant of this model was then used, which added random responding: on a certain proportion of trials (p), the model responded S or L with equal probability regardless of the stimulus duration presented. The value of this parameter decreased with increasing age in the 1/4-s condition, taking values of 0.2, 0.1, and 0.01 for the three age groups. In the 2/8-s condition, the values were 0.1, 0.15, and 0.01, respectively. The other parameters showed a pattern similar to that found when the basic model was applied; in particular, memory/timing variance was higher in the 3- and 5-year-olds than in the 8-year-olds.

Droit-Volet, Clément, and Wearden (2001) conducted a similar study, again employing children 3, 5, and 8 years of age, but this time on a temporal generalization task. The standard durations were either 4 or 8 s, and comparison durations were spaced at either 1-s intervals (4 s) or 2-s intervals (8 s) around them. The stimuli whose durations were to be judged were blue circles presented in the centre of the computer screen.

The gradients obtained from the 8-year-olds with both standard duration values were "adult-like" in the sense that they were peaked at the standard and were skewed to the right. For the 3- and 5-year-olds, on the other hand, the gradients were either symmetrical or slightly skewed to the left. The generalization gradients for the 3- and 5-year-olds were also considerably flatter than those for the 8-year-olds, and in the children, a high proportion of YES responses occurred to stimuli which were remote from the standard. For example, the 3-year-olds produced about 25 % YES responses at the shortest stimuli (1 and 2 s) for the different standards. The temporal generalization gradients from each group superimposed well when comparison durations from the conditions compared were expressed as a proportion of the standard for that condition (see their Fig. 3, p. 297).

Data were modelled with a variant of the MCG model developed by Wearden (1992) and discussed in Chap. 4. In addition to the two normal parameters (timing/memory variability, c, and threshold, b), a random responding parameter (p) was added. As in the bisection simulation above, the two possible responses, YES and NO, occurred with equal probability with probability p, regardless of the stimulus duration presented. In addition, following McCormack et al. (1999), a "distortion" parameter, k, was added, which multiplied the standard value.

Memory/timing variability was greater in the 3- and 5-year-olds than in the 8-year-olds, whereas thresholds differed only slightly, being higher (i.e. less conservative) in the older children. The random responding parameter yielded considerably higher values in the 3- and 5-year-olds (between 0.52 and 0.22) than in the 8-year-olds (0.12). In addition, distortion values were smaller on average (between 0.83 and 0.9) in the 3- and 5-year-olds than in the 8-year-olds (0.9 and 1.0).

In the bisection case (Droit-Volet & Wearden, 2001), adding a random responding parameter improved the fit of the model, although it was reasonably well adjusted to the data without it. In the temporal generalization case, however, the random responding parameter was essential, as without it the high proportions of YES responses at comparison durations remote from the standard could not be simulated.

These three earliest applications of SET or an SET-like model to data from children laid the groundwork for later research. All three found that timing/memory variability decreased on both temporal generalization and bisection tasks as children aged, that is, the precision of their time representations improved. The two studies able to test it also found superimposition, suggesting the manifestation of scalar properties of time even in children as young as 3 years. For temporal generalization, increasing age resulted in less stringent criteria for responding YES, and a "distortion" parameter was needed to model the generalization gradients, particularly those obtained from the youngest children.

The first of these findings, that temporal precision increases with age until around 10 years, at which time values approach those obtained from student-age adults, has received consistent confirmation in subsequent work, as material later in this chapter will show. The necessity of proposing "distortion" in the youngest children's temporal generalization gradients, however, has been more controversial.

Droit-Volet (2002) tested children 3, 5, and 8 years of age on a temporal generalization task with standards of 0.4 s and 4 s. Temporal generalization gradients increased in steepness with age, and showed good superimposition. The author then modelled the data using the same model as that in Droit-Volet et al. (2001), but without distortion, and found that it fitted reasonably well, suggesting that "memory distortion" was not necessary to account for data even from the youngest children. In contrast, McCormack et al. (2004) found evidence for "distortion." In an experiment using a 500-ms standard and a generalization technique with children 6 and 10 years of age and adults, they replicated the basic result of McCormack et al. (1999), namely that gradients obtained from the youngest children were skewed to the left and needed "distortion" to model them. A second experiment used a "repeated standard" method where a 500-ms standard tone was presented first on each trial, followed by a comparison tone whose duration was some multiple of the standard. Thus the standard was refreshed on each trial, and did not need to be retrieved from a reference memory store. In this case, gradients from the 6- and 10-year-olds showed adult-like rightward skewing, although the steepness of the gradients increased with age, as in Droit-Volet et al. (2001). The obvious conclusion from this work is that memory distortion arises because of the memory load imposed by the need to retrieve the standard in the "normal" temporal generalization procedure.

The suggestion that young children may have a temporal reference memory qualitatively different from that of adults, in that standard durations are stored as systematically shorter than they are, is an intriguing one; however, data remain inconclusive, as shown above, although more data support "distortion" than argue against it. Nevertheless, the issue remains unresolved, as most subsequent studies of timing in children within the SET framework have used bisection rather than temporal generalization, and this was modelled by McCormack et al. (1999) and Droit-Volet and Wearden (2001) without distortion. This suggests an inconsistency in the treatment of the two tasks from a psychological point of view. If young children remember the standards in temporal generalization as shorter than they are, why do they not also remember the standards in bisection as shorter? Using computer simulation, Wearden and Jones (2013) show that such an effect would be easily detectable in data if the distortion were considerable. One possibility, discussed earlier (Chap. 4), is that the children, like the adults studied by Wearden and Ferrara (1995), are not actually using the standard durations in bisection as reference for their performance, and instead are using some sort of constructed criterion value, for example, the mean of all the stimulus durations presented.

In a rare study in which bisection was not used, McCormack, Wearden, Smith, and Brown (2005) investigated children's performance on two variants of the normal temporal generalization procedure. One was the "episodic" method developed by Wearden and Bray (2001), discussed in

Chap. 4. Here, two stimuli are presented on the trial, one with a random value and the other some multiple (e.g. from 1.75 to 0.25) of it. The participant's task is to judge whether the two stimuli have the same duration. With this procedure, generalization gradients can be produced even though there are no "standards," and thus presumably no involvement of reference memory. Another variant was the "repeated standard" task, discussed above. Two duration ranges (all < 1 s) were used to enable a test of superimposition, and performance of children aged 5 and 10 years was contrasted with that of adults. Generalization gradients increased in steepness with age on the episodic task, and this was also true on the repeated standard task, although the performance improvement occurred only between 5 and 10 years of age. Superimposition was found for all age groups with both tasks. The generalization gradients in both the episodic and repeated standard cases were shifted to the right in 10-year-olds and adults, but were more symmetrical in the 5-year-olds (cf. Droit-Volet et al., 2001), although not clearly left-shifted. This is consistent with McCormack et al.'s (2004) notion of "distortion" as a property of reference memory, as neither episodic nor repeated standard generalization necessitate the use of reference memory, although it could be used in the repeated standard task, as individuals may store the first stimulus on the trial in reference memory.

Provasi, Rattat and Droit-Volet, (2010) extended the range of applicability of SET-consistent models down the developmental scale by investigating bisection performance in 4-month-old infants, using direction of gaze as a response measure. With a *Short/Long* standard pair of auditory stimuli 500 and 1500 ms in duration, orderly psychophysical functions with near-geometric mean bisection were obtained. Modelling suggested that timing/memory variability values were close to those obtained in animals, and that a considerable degree of random responding contaminated the psychophysical functions.

A number of studies with children which can be subsumed within the SET framework have used a developmental context to explore "classic" effects in time perception, such as the type of stimulus used, as discussed in Chap. 3. For example, Droit-Volet, Tourret, and Wearden (2004) showed that, like adults, children aged 5 and 8 years judged auditory stimuli as lasting longer than visual stimuli of the same real duration (for a demonstration of this effect with adults, see Wearden et al., 1998; Wearden, Todd, & Jones, 2006). In a similar vein, Droit-Volet (2008) used children 5 and 8 years of age, along with an adult comparison group, to explore the "filled-duration illusion" (Wearden, Norton, Martin, & Montford-Bebb, 2007) using a bisection technique and auditory stimuli. The filled stimuli were continuous tones, and the empty ones the same interval started and ended with short click-like stimuli. When tested in a within-subject design, where participants were exposed to both types of stimuli, the children showed this classic effect, possibly in a more marked manner than adults, although sometimes the effect was reversed at the longest durations (see Droit-Volet, 2008, Figs. 1 and 3, for example). When a between-subject design was used, such that filled intervals were compared with other filled intervals and empty intervals with other empty ones, the effect disappeared in all groups.

Although SET-inspired studies seem very distinct from earlier work influenced by Piaget, Droit-Volet, Clément, and Fayol (2003) conducted a bisection experiment with a distinctly Piagetian flavour. In their Experiment 1, children 5 and 8 years of age and adults were initially trained to distinguish *Short* and *Long* standards 2 and 8 s in duration. The stimuli were presented in the form of successive blue squares, each 1 s long, so there were two stimuli presented for the *Short* standard and eight for the *Long*. Following training, the participants were tested with two types of stimulus sequences. One of these (time varying) held the number of stimuli constant while varying the total duration of the sequence between 2 and 8 s. In the second condition (number varying), the sequence duration was constant at 4 s, but the number of stimuli presented varied from two to eight. As the basis of the task was the total duration of the stimulus, the ideal strategy would have been to ignore the number of stimuli in the number-varying condition, but the 5-year-olds were unable to do this, particularly in one of the conditions when counting was prevented. Here, the youngest children produced the same psychophysical functions for both time and number variation, as though these were impossible to distinguish. Older children and adults were less affected by variation in number when time was held constant. In a second experiment, the basis of the task was the *number* of stimuli presented, not total duration, but the procedure was otherwise the same as the first.

Now, the numerical bisection performance of even the youngest children was little affected by the irrelevant stimulus dimension, time. This study is obviously reminiscent of the Piagetian and neo-Piagetian concepts that time is a more difficult dimension of a task to extract, particularly for younger children, than other aspects of the experimental situation such as spatial displacement or, in this case, number. In the first study, "larger number" equalled "longer time," although the reverse was not true, with numerical bisection little affected by the time taken to present the stimulus sequence.

Further evidence of the difficulty of time discrimination relative to other types of judgements comes from Droit-Volet, Clément, and Fayol (2008). In their first experiment, bisection tasks were carried out by children aged 5 and 8 years and adults, where the basis of the task was either time, number, or length. For time discrimination, the stimuli were blue squares, with *Short* and *Long* standards of 4 and 10 s or 8 and 20 s, in different conditions. For the number bisection, the standards were the number of small blue circles, with *Few/Many* values of 4/10 or 8/20. In the length discrimination, the stimulus was a blue line, with standards of 4/10 or 8/20 cm. After initial discrimination training, testing was conducted with a range of values presented (e.g. 4–10 s, 4–10 circles, 4–10 cm).

Psychophysical functions for all groups were flatter for time discrimination than for number or length, with a particularly marked effect in the 5-year-olds. In a second experiment, the same modalities were used, but this time the stimulus values were presented "sequentially": the stimuli in the time case consisted of a sequence of durations which participants were required to add together, and number and length were also presented in this sequential fashion. In this case, the differences between time, number, and length disappeared in all groups (although psychophysical functions became steeper with age). These results suggest that the difficulty experienced by individuals—particularly young children—with time discrimination is that time is necessarily "sequential," in that the duration of a stimulus necessarily unfolds over time, whereas number and length discrimination in almost all cases is based on the immediate perception of the stimulus presented. Presumably, the sequential nature of duration imposes loads on children's attentional and memory capacity, whereas the usual non-sequential presentation of number and length is easier. Thus, although this experiment supports the traditional Piagetian position that judgements about time are especially difficult, the reason may be related to differences in more general cognitive processes such as attention and memory in younger children rather than difficulties in processing time per se.

Another study with a Piagetian flavour, albeit not conducted within the SET framework, was carried out by Droit-Volet and Rattat (1999). Children 3.5 and 5 years of age were trained to squeeze a bulb for 5 s. A day later they were asked to produce the same duration, with different groups using either the same action or a different response: holding down a button. The 5-year-olds transferred well to the new task, whereas the younger group failed. A second experiment found the same result when the transfer test was conducted immediately after training. A third experiment used more elaborate instructions intended to overcome the possible issue that the young children had failed to understand the nature of the task. While this somewhat lengthened the durations reproduced, transfer from one response to another still eluded the younger children. Overall, the pattern of results suggests that duration was linked to non-temporal aspects of the action, such as force, and that children have difficulty dissociating temporal and non-temporal aspects of a task, a traditional Piagetian theme.

Following on the idea that factors like memory might be important in explaining developmental differences in the performance of timing tasks, a number of studies have explored memory for duration in a developmental context. Rattat and Droit-Volet (2005) used children aged 3, 5, and 8 years on a bisection task with blue circles of 2 and 8 s as stimuli. After initial training, children were tested either immediately or after 15 min of a question-and-answer game intended to produce interference with the memory of the standards. The interference flattened psychophysical functions in the 3- and 5-year-olds, but had no effect in the 8-year-olds, indicating greater fragility of memory of the standards in the younger children. A second experiment used 5- and 8-year-olds on the same task, this time without an interfering task, but with either immediate testing or testing delayed for 15 min or 24 h. Consistent with the concept of greater fragility of temporal reference memory, the 5-year-olds showed flatter psychophysical functions after the two delays, particularly after 24 h, whereas the older children were less affected.

However, greater persistence of memory for duration in young children over longer periods was found using a different method in Rattat and Droit-Volet (2007). Here, children 3 and 5 years of age were trained to depress a button for 5 s by imitating the experimenter's action. Retention delays of 1, 24, or 48 h were introduced before testing. Children in both groups showed evidence of retention of the duration they had learned, even after 48 h. In a second experiment, 5-year-olds learned a 5-s duration by imitation, then were tested after 6 days, 6 weeks, or 6 months. While performance deteriorated as the retention interval increased, there was evidence of retention even after 6 months. These two experiments seem to suggest that memory of temporal standards is susceptible to both interference and forgetting effects, and that younger children are affected to a greater degree than older children, yet the degree of performance deterioration is dependent on the task, with evidence of retention of learned time intervals strongest when a motor response is learned by imitation.

Short-term retention of duration was also studied in a developmental context by Droit-Volet, Wearden, and Delgado-Yonger (2007). The task was a type of "episodic" temporal generalization (Wearden & Bray, 2001). As typical in studies of short-term memory for duration (such as Wearden & Ferrara, 1993, discussed in Chap. 3), the details of the procedure were complicated, but the basic method involved children 5 and 8 years of age and adults, who received two blue circles, with a delay of 500 ms or 5 or 10 s between them. One of the stimuli had a randomly generated duration (between 1200 and 2000 ms), and the other was some multiple (between 0.25 and 1.75) of it. The task was to judge whether the stimuli were equal in duration (assessed by a YES or NO response). The resulting temporal generalization gradients increased in steepness with increasing age (cf. Droit-Volet & Wearden, 2001), but the focus of interest in the experiment was the effect of the delay between the stimuli to be judged. All groups behaved as though the first stimulus was shorter than the second, when they were in fact the same, although the effect was greatest in the youngest children. Another obtained effect was a flattening of the generalization gradients with increasing delay between the stimuli, once again with the effects more marked in the 5-year-olds. Data were accounted for using a rather complex model based on SET-consistent ideas: modelling showed that memory/timing variability increased in all groups (but most in the youngest children) as the inter-stimulus delay increased.

Overall, these studies of short-term and longer-term memory for duration in children show a consistent picture. Memories become more stable, thus more resistant to interference and imposition of various sorts of delays, as children age, suggesting that at least part of the reason for developmental changes in timing performance may be changes in memory with age.

In addition to memory changes with age, children of different ages may react differently to attentional manipulations associated with timing tasks. One of the first studies of attention and timing in children was conducted by Arlin (1986). In a first experiment, Arlin asked children 11 years of age to reproduce the duration of a sequence of pictures. In different conditions, these received two different levels of processing (shallow and deep), and different numbers of pictures, either four or eight, were presented. Deeper processing (in this case deciding whether each picture represented a living or non-living thing) shortened reproduced durations, whereas increasing the number of pictures increased perceived durations. A second experiment used a similar procedure with children approximately 6, 8, 10, and 12 years of age, again with deep and shallow processing, but this time with three or six pictures. The interval duration was 9 s. Once again, deep processing produced shorter reproductions than shallow, and the duration reproduced increased with the number of pictures presented. In general, the 9-s interval was markedly underestimated in all conditions, but average reproduction length increased with age.

Zakay (1992) carried out a study (his Experiment 2) which drew inspiration both from Piagetian time psychology and from attentional models of timing (Chap. 5). Children between 7 and 9 years of age were required to reproduce the duration during which light-bulbs were illuminated, using lengths of 6 and 10 s. The bulbs were either large and high-intensity or small and low-intensity, and a distracting noisy toy was sometimes activated during illumination of the bulb. The bigger, brighter bulb was judged as having lasted longer than the smaller, dimmer one, and the presence of an attentional distractor shortened reproductions.

The presentation of potential distractors, however, does not always shorten children's time judgements. Gautier and Droit-Volet (2002a) showed a striking difference with age in the effects of distractor stimuli during a timing task. Two groups of children were used, aged 5.6 and 8 years on average, and the task was bisection, with 2/8-s *S/L* standards and test stimuli spaced in 1-s steps between these values. The stimulus was a red circle in the centre of a computer screen, which was presented either alone or accompanied by a series of black geometric figures which changed every 250 ms, intended as a distractor. In the 8-year-olds, the presence of the distractor stimuli shifted the psychophysical function to the right, that is, the distractors made the target stimuli generally seem *shorter* than without the distractor. In the 5.6-year-olds, the distractor stimuli had the opposite effect: the psychophysical functions were shifted to the left, indicating that the stimuli were judged as *longer* in the distractor condition. This result from the younger children has echoes of the "more = more time" message of Piagetian time psychology; presumably, the increased sensory input involved when distractors are present results in durations judged as longer than when only red circles are presented. The older children, in contrast, show a more adult-like distraction effect. The result from the younger group is also reminiscent of Thomas and Weaver's (1975) assertion that temporal and non-temporal information can combine to lengthen duration judgements. However, in contrast to the results of their experiment using bisection, Gautier and Droit-Volet (2002b) found that reproductions of durations of 6 and 12 s in both 5- and 8-year-olds were shortened when a picture-naming task was conducted at the same time as the reproduction, in comparison to intervals reproduced without the secondary task.

Droit-Volet (2003) examined what she called "alerting attention." In this study, children aged 3, 5, and 8 years carried out a bisection task with *S/L* standard pairs of 0.5/2 s or 1/4 s. The stimuli were visual. In half of the test trials, an auditory click occurred 1.5 s before the visual stimuli were presented, and in the other half, the click was absent. The click shifted psychophysical functions to the left in all age groups (i.e. stimuli were judged as longer after the click). There was no overall effect of age with respect to the clicks, but the steepness of the psychophysical functions increased with age, as is typical in bisection in children. The leftward shift of the psychophysical function was the same for the 0.5/2-s and 1/4-s bisection conditions. Such an additive effect is consistent with the notion that the alerting attention manipulation altered switch operations in the SET system, rather than increasing pacemaker speed by increasing arousal, which would have produced larger effects at the longer duration range (see the appendix to Chap. 3 for discussion of potential differences between pacemaker and switch effects).

Finally, studies attempting to predict children's performance on timing tasks using measures of more general cognitive processes (e.g. Zélanti and Droit-Volet, 2011) are clearly relevant to the question of the extent to which differences in children's timing performance compared with that of adults is dependent on some factor specifically related to timing or to differences in other psychological processes. This work was reviewed in Chap. 5.

Overall, SET-inspired studies of the development of timing performance, the vast majority of them published by Sylvie Droit-Volet and colleagues, have led to considerable advances in our understanding of which features of timing change developmentally, as well as insight into why this might occur. A virtually universal finding is that performance on timing tasks improves in terms of timing sensitivity until 10 or 12 years of age, when near adult-like performance is observed. Some evidence suggests qualitative differences in how temporal standards are remembered in the youngest children compared with older children and adults, as well as the decreasing fragility of memory for time as children age. When data can test it, something close to scalar timing is obtained, even from children as young as 3 or 5 years. Later work has focussed on the involvement of various "general" psychological processes, such as short-term or working memory, different types of attention, or information-processing speed, on performance of timing tasks. Data clearly suggest that influences exist between memory, attention, and timing, but exactly which processes contribute to different timing tasks, and their degree of contribution, remains unresolved.

Summary

Piaget's book *The Child's Conception of Time* suggests that young children's notions of time are very different from those of adults. Young children may have particular difficulty distinguishing duration from other aspects of the experimental situation, may not agree that events which start and stop at the same time last for the same length of time, and may not recognize that a shorter interval of time is necessarily included within a longer one. Later researchers in the Piagetian tradition have generally confirmed Piaget's results. With young children, "more = more time," so comparisons of greater distance traversed, greater speed, or greater display brightness may all yield erroneous judgements about time. In addition, younger children may find it difficult to coordinate the start and end points of sequences, often focussing on one or the other alone. Certain research on the potentially complex relations between speed, distance, and time judgements suggests that a full "scientific" understanding of these relations may not occur until young adulthood.

Studies using SET or SET-type models have been performed extensively with children as young as 3 years (and in one case with infants). Bisection and temporal generalization methods have been most common. In general, data support the idea that precision of timing improves with age until 10 or 12 years. Children 3 or 5 years of age may have temporal memories where standard durations are stored as slightly shorter than they are, although this is a less certain finding. Young children may also perform poorly on timing tasks because of attentional problems, or lower general cognitive capacity than that in older children and adults. As in Piagetian studies, some work suggests that the dimension of duration is more difficult to extract from experimental settings than the dimension of space, reminiscent of ideas in Piagetian work.

8

Timing and Ageing

At the other end of the developmental spectrum, post-young adult development, there has been considerable interest in the possibility of significant changes in time perception and time experience as people age. Why should any sort of changes be expected? One characteristic of normal ageing, the slowing of information processing, might be expected to affect performance on timing tasks. More specifically, slowing metabolism with age may produce a "slowing down" of the pacemaker of the internal clock, which as we have seen has been a popular underlying theoretical principle for explaining human timing. In addition, increasing age brings with it changes in memory, attention, and other cognitive processes; thus older people have lower attentional and memory capacity and poorer control of their executive functions than younger individuals (Hedden & Gabrieli, 2004). These changes in cognitive function with age naturally are linked to changes in the brain (Reuter-Lorenz & Lustig, 2005). All these considerations suggest that changes in performance on timing tasks might well be expected as people age.

In this chapter, I will consider two types of studies that have been conducted. The overwhelming majority of these involve the performance of older individuals compared with student-age control groups using

J. Wearden, *The Psychology of Time Perception*,
DOI 10.1057/978-1-137-40883-9_8

laboratory tasks of timing. However, regardless of the results of studies of standard timing tasks, a common impression is that changes in time experience occur as people age—the cliché that older people perceive time in their daily lives as passing more and more quickly, for example: "Christmas comes round quicker every year." There are difficult questions about just what such statements actually mean, and some studies have looked specifically at potential changes in what one might call "time experience" with increasing age.

Ageing and Performance on Standard Tasks of Timing

Numerous studies, many reviewed by Block, Zakay, and Hancock (1998), have examined the performance of older people (usually 60+) on tasks such as verbal estimation of duration and interval production and reproduction. Two clear messages emerged from Block et al.'s meta-analysis. The first was that what they called the *duration ratio*, time judgement divided by objective time, tended to be larger among older groups. The second was that the variability in timing judgements was almost always greater in an older group than a younger one; for example, a measure of variability such as the coefficient of variation was almost always higher in older individuals than in student-age comparison groups.

Some of the striking effects of ageing on performance of timing tasks, however, may have causes that are less interesting than they seem at first sight. For example, Craik and Hay (1999) reported significant differences in performance between an older (72.2 years) and younger (22.2 years) group in the production and verbal estimation of a range of intervals from 30 to 120 s. Intervals produced by the older group were all longer than those of the younger group. However, production even in the younger group was not accurate, with intervals produced around 25 s longer than they were in reality. In contrast, both groups underestimated all intervals, with underestimation particularly marked in the older group. Once again, however, neither group was particularly accurate on average. For the 120-s interval, for example, the younger group estimated it at around

75 s, the older group around 35 s, judging from data shown in Craik and Hay's Fig. 2. These results are striking, but perhaps they have a rather mundane explanation. With the intervals used in this study, participants would almost certainly employ chronometric counting, which was not prevented or controlled in the experiment, to guide their performance. The effects obtained may thus be interpretable simply in terms of a slow rate of counting in the older group, perhaps occasioned by motor slowing or slowing of "internal tempo" (Vanneste, Pouthas, & Wearden, 2001). Slower counting in the older group would result in longer productions, as more clock time would be needed to count up to any particular target value, and in shorter estimations, as during any fixed time period, a lower count would occur in the older group than the younger one. Therefore, even quite dramatic effects like those obtained by Craik and Hay (1999) may not imply any radical change with age in the perception of time.

What does the performance of older people look like in situations where counting is not helpful because the intervals used are too short? Wearden, Wearden, and Rabbitt (1997) conducted a study of this sort, also using SET-based modelling, in an effort to uncover the theoretical processes underlying performance changes with age. The study used two age groups, 60–70 years and 70–80 years. Participants were sampled from a large subject panel in which individuals' IQ measures were known, so the usual confound between IQ and age (i.e. a decline in IQ with age) was avoided. This meant that IQ values on average did not differ between the two age groups, and thus data could be analysed in terms of age (with IQ constant) and IQ (with age constant). Four timing tasks were conducted: temporal generalization, temporal bisection, a discrimination task, and an interval production task.

For the temporal generalization task, the standard duration was a 400-ms tone, with the comparison spaced in 100-ms steps around the standard. Generalization gradients peaked at the standard in all between-group comparisons, and were rightward-skewed, as was found in student-age participants (e.g. Wearden, 1992), but gradients were flatter in the 70–80-year-olds than in the younger group. The flattening was particularly marked when data were analysed by IQ (with the group divided into the highest third, middle third, and lowest third of the IQ values). Fits of the MCG model (Chap. 4) showed that the main effect was on timing/memory variability (the *c* parameter), which revealed an increase with increasing age and a more marked increase with decreasing IQ.

For the bisection task, the *Short* and *Long* standards were tones of 200 and 800 ms, with comparisons spaced linearly in 100-ms steps between them. In contrast to the effects of age and IQ on temporal generalization, there were no significant effects of either age or IQ on bisection. A comparison with previously published data from student-age participants also showed that the bisection performance of the older individuals was very similar to that of individuals around 50 years younger on average.

The finding of an effect of age on temporal generalization but not bisection was replicated by McCormack, Brown, Maylor, Darby, and Green (1999), albeit in a study where IQ was not controlled. Their modelling, which as noted in Chap. 7 used a model similar to SET, found increased "noise" (i.e. timing/memory variance) with increasing age, a result very similar to that obtained by Wearden et al. (1997). In addition to temporal generalization and bisection, Wearden et al. used a discrimination task, where participants had to decide whether a tone of variable duration was the same as a 600-ms standard tone presented on the trial. If the decision was that the durations of the stimuli were different, the variable tone became closer in duration to the standard on the next trial. Different conditions varied whether the standard was presented first or second on the trial, and whether the comparison duration was initially longer or shorter than the standard. Lower-IQ participants showed poorer discrimination in general, but there was no overall effect of age. However, in the condition in which the standard came first and the comparison was initially shorter, the 70–79-year-old participants discriminated the stimuli more poorly than the 60–69-year-olds.

The final task involved the production of a 1-s interval between two presses on the spacebar of a computer keyboard. Initially, no feedback was given regarding performance accuracy; feedback was then introduced, and was subsequently discontinued. Without feedback, the lower-IQ and younger groups showed higher mean productions than the middle- and higher-IQ and older groups. Feedback eliminated any between-group differences, and this absence of difference persisted when feedback was withdrawn. When the coefficient of variation of productions was used as a measure, there were no age effects, but there was greater variation in productions of lower-IQ participants than those of higher-IQ individuals.

In summary, then, Wearden et al.'s (1997) study found effects of age and IQ—with the latter usually clearer—on some timing tasks (temporal generalization, discrimination, and variability of interval production) but not on others. The absence of an effect on temporal bisection was particularly striking, as Wearden et al. also showed that bisection performance of their participants was very close to that of student groups on average about 50 years younger, with data taken from bisection studies with students.

Overall, there was no suggestion that the performance of older participants in the Wearden et al. (1997) study was in any way abnormal. For example, on temporal generalization, gradients peaked at the standard and were rightward-skewed, and were well fitted by the MCG model as well—all characteristics of the data from student-age participants. Other work suggests that the performance of older individuals is very similar to that of students on a range of tasks. For example, if we consider the control groups (average age 65 years) from Wearden et al. (2008, 2009), we see performance on temporal generalization and bisection resembling that of students. Older participants showed not only typical verbal estimation performance, but also the typical effect that trains of clicks preceding stimuli made them appear longer (Wearden et al., 2009), and older participants showed the Wearden and Ferrara (1993) "subjective shortening" effect when tested on a memory-for-duration paradigm in Wearden et al. (2008). Figure 8.1 shows the performance of the control group from Wearden et al. (2008; average age close to 66 years) on bisection, temporal generalization, and verbal estimation tasks, all using durations too short to make counting useful.

These findings suggest that while performance variability may be higher in older individuals than those of student age, there are no marked differences in the type of performance obtained, particularly in situations where participants receive feedback or can calibrate their performance against some sort of standard. What, then, about the idea that the pacemaker of a proposed internal clock might slow with age? Perhaps the effects are smaller than might first be thought. One issue is recalibration. Suppose two individuals have different pacemaker speeds, so for one the

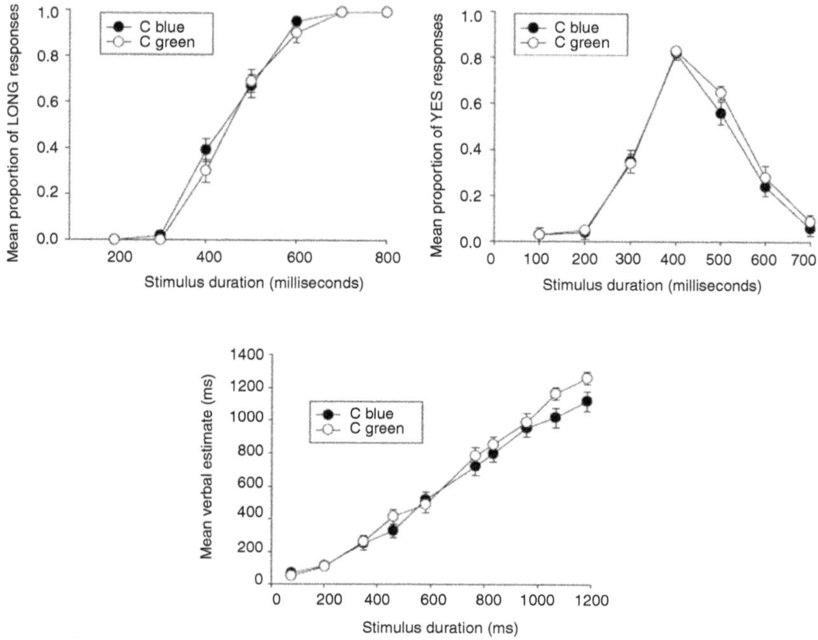

Fig. 8.1 *Upper left panel*. Bisection performance from the control group in Wearden et al. (2008). *Upper right panel*: Temporal generalization gradients from the control group in Wearden et al. (2008). *Lower panel*: Mean verbal estimates plotted against stimulus duration from the control group in Wearden et al. (2008). The green and blue indicate two different sessions, which were counterbalanced

pacemaker "ticks" at *m* per second, and for another at *n* per second. In a task such as interval production with feedback, the two individuals might produce the same mean value if required to produce exactly one second: the timing system of one person does this using *m* ticks of the clock, whereas the other uses *n*.

Some direct evidence that calibration might obscure potential differences in pacemaker speed between older and younger individuals comes from Vanneste et al. (2001). An older group and student-age group received two tasks. One was a measure of "internal tempo" (Denner, Wapner & Werner, 1964). Here, people are asked to tap for a short period at a speed which feels comfortable to them, and the tap rate defines internal tempo.

The older group had a significantly slower internal tempo. One possible interpretation of internal tempo, albeit one for which the evidence is not strong, is that it reflects "clock speed," so if this is assumed, the older participants in Vanneste et al.'s study had a slower internal clock pacemaker than the younger participants. The other task used was the continuation tapping task devised by Wing and Kristofferson (1973). Here, participants are initially synchronized to a series of periodic sounds; then the sounds stop, but the participants are required to continue tapping at the same rate. In Vanneste et al.'s experiment, the synchronization intervals were 300, 400, 500, 600, and 700 ms. In the continuation tapping task, there were no age-related differences, and both groups on average tracked the target interval nearly perfectly. This suggests that the calibration of between-tap intervals provided in the continuation tapping task enabled both the older and younger groups to synchronize their taps almost perfectly, despite a proposed difference in internal clock speed (see Fig. 8.2).

If two individuals or groups do differ in pacemaker speed, then feedback or calibration may make mean differences in their timing behaviour impossible to detect, but there may still be other means of detecting clock speed changes between groups in terms of performance variability. The mathematics of internal clocks shows that slower clocks will produce time representations with greater variation than faster clocks. In other words, for any particular time interval, the number of accumulated ticks will be more variable (e.g. as a proportion of the mean number of ticks accumulated) with a slow clock than with a faster one. As noted above, increased performance variability with age was a notable feature of results reviewed in Block et al. (1998), consistent with the notion that older individuals have slower pacemakers, provided of course that pacemaker speed is an important contribution to the variability of the timing task overall.

Over the last decade, there has been growing interest in the contribution of more general psychological processes, such as attention and memory, to timing tasks, and some of this work was discussed in Chap. 6. This type of research could be fruitfully applied to the study of ageing and timing, and evidence already exists that cognitive changes such as those in attentional capacity might explain all or part of the difference in timing performance between older people and those in their early 20s.

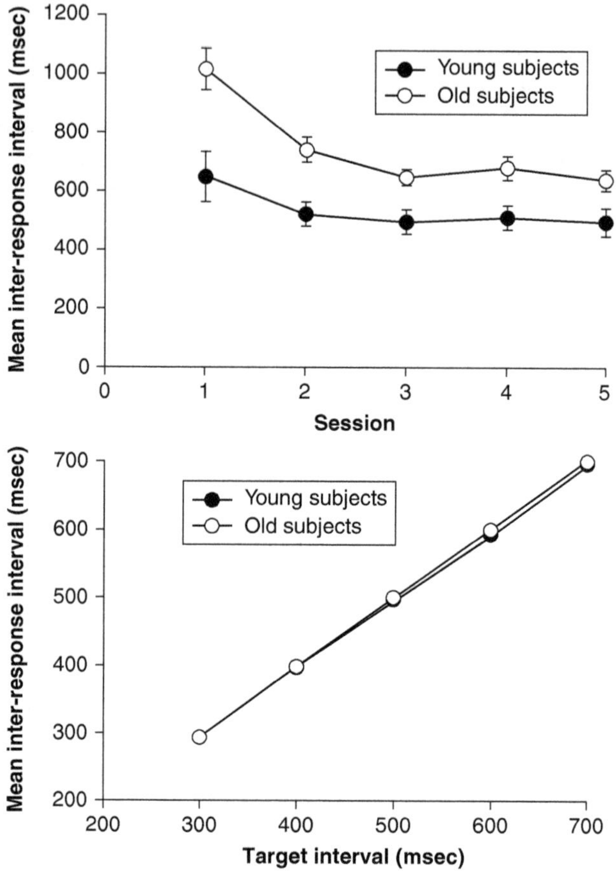

Fig. 8.2 Data from the young and elderly participants from Vanneste et al. (2001). *Upper panel*: Performance on the "internal tempo" task. The mean inter-response times from the two groups are shown over five sessions. *Lower panel*: Performance on the continuation tapping task. The mean inter-response times are plotted against target interval in ms

Lustig and Meck (2001), for example, conducted a bisection task with 3-s and 6-s *Short/Long* standards, and with black squares or tones as the stimuli. These were sometimes presented alone, but at other times both the auditory and visual stimulus were presented on the same trial with different durations, onsets, and offsets. Performance of a group of

69-year-olds was compared with that of 20-year-olds. Both groups performed more poorly in terms of temporal sensitivity in the compound condition than in the single condition, but the older group seemed to have particular difficulty in the compound case. A similar study by McAuley, Miller, Wang, and Pang (2010) used the human peak procedure method (Rakitin et al., 1998), with auditory or visual signals presented either alone or in combination. As in Lustig and Meck's results, the older group (60–74 years) showed lower timing sensitivity in the form of greater spread of the peak function in the compound condition. This was also found for the younger group (18–39 years) but to a much lesser extent. McAuley et al. then correlated measures of working memory span and Stroop interference with timing task variability. Age was positively correlated with variability in the compound task case, as was working memory span. These two results together suggest that part of the reason for age-related differences in timing performance may come from differences in psychological processes like memory and attention, a position similar to that held by Zélanti and Droit-Volet (2011) with respect to changes in timing behaviour in children.

Implicating non-timing psychological processes in the performance of timing tasks may also address the general question of why some timing procedures seem to be affected by ageing more than others, in particular the puzzling failure to find effects of age on bisection, when they are found on temporal generalization with the same participants in the same experimental setting (Wearden et al., 1997; replicated by McCormack et al., 1999). Ogden, Wearden, and Montgomery (2014) in a study using student participants discussed in Chap. 6, looked at correlations between aspects of executive function (switching between tasks, inhibiting unwanted responses, and accessing memory) and performance on three timing tasks (temporal generalization, reproduction, and verbal estimation). They found that the different executive functions made different contributions to different timing tasks, but more generally, that performance on different tasks may be dependent on different "general" psychological components (e.g. Droit-Volet, Wearden, & Zélanti, 2015), and this might be part of the reason that effects of ageing seem to depend on the timing task used.

Time Experience in Older People

In casual conversation, most older people seem to agree that there are changes in time experience as people age—saying, for example, that "Christmas comes around quicker every year" or that "time passes more quickly as you get older." There are obvious questions about what such statements actually mean: an elderly person knows that the interval between successive Christmases remains the same, but somehow it "feels" shorter than it used to, or used to be imagined to feel. In fact, most of the experimental data supporting the notion of some sort of speeding up of the feeling of passage of time with increasing age comes from a method introduced by Lemlich (1975).

Lemlich asked individuals to judge how fast time seemed to pass when they were half or one-quarter of their current age. Most people—even those in their 20s—said that time passed more slowly when they were younger. The average rating of passage of time at a quarter of their present age was about half as fast as at their current age. Given that this was described even by young people, it is not clear that it is a "developmental" effect at all, as it is definitely not dependent on being elderly. One obvious problem is the question of what people are actually reporting here. Even if a person is only 20, can they really access what time passage felt like when they were 5 years old? Nisbett and Wilson (1977), in their influential discussion of the role of introspective reports in cognitive psychology, cautioned against relying on reports of processes to which a person could not reasonably be assumed to have access, whereas they regarded reports of accessible processes as more reliable. The fact that people agree about something does not, in itself, validate the thing about which there is agreement, and a clear example of this comes from another study using the Lemlich method, that of Joubert (1990). Here, participants were asked how fast time would pass when they were older than their current age, and reliable agreement about future speeding up was obtained, although this judgement obviously cannot be based on direct experience.

Suppose that people are asked more systematic questions about the passage of time, with different age groups used. What results are obtained? Two very similar studies addressed this question. Wittmann and Lehnhoff

(2005) studied 499 participants in different age bands, with the youngest 14–19 years and the oldest 80+. Individuals were asked questions about the rate of time passage (on a scale from −2 to +2), as well as questions about time pressure. Participants were asked questions about the present ("How fast does time usually pass for you? How fast do you expect the next hour to pass?") and questions about the past ("How fast did the previous week/month/year/10 years pass for you? How fast did your childhood (<12 years) go by?" and so on). Asked about the present time, all groups gave a mean above 0, so this was judged as slightly fast, but there were virtually no effects of age. When asked about past time, only the question of how fast the previous 10 years had passed produced clear age effects, passing more quickly in those 40 and over than in younger individuals, but with the fastest speed in the 50–59 and 60–69 groups.

Friedman and Janssen (2010) conducted a very similar study with 1766 Dutch participants, and obtained virtually identical results, with only the "previous 10 years" question showing a clear effect of age. This was actually their second experiment. In the first, they tested a possible explanation for age-related effects on time experience called "forward telescoping." This refers to a tendency to underestimate how long ago certain events occurred. If a person does this, when the true date is then revealed, an impression that time is passing quickly might be gained. For example, if a person judges that some event occurred 2 years ago, and then is told that it was actually 5 years ago, they may conclude that the 5 years passed exceptionally quickly. If older individuals show more forward telescoping than younger individuals, this may explain the age-related effects on the subjective speed of passage of time.

Friedman and Janssen used a New Zealand-based sample of younger and older people and asked them to estimate how long ago 12 news events occurred. The participants then received information about the dates of the six events they best remembered, but for half of the participants in both age groups this information was false, with some items described as one-third again more or less old than they really were. Subsequently, items from the Wittmann and Lehnhoff questionnaire were administered. Considering the estimates of how long ago the events occurred, there were no significant differences between the older and younger groups when deviation from the true date was taken without regard to sign. When the difference was signed, the younger group underestimated the time elapsed from the event to a greater degree than did the older group.

The forward telescoping manipulation did have some effect, as those who were told their estimates were too long produced shorter estimates than those told their estimates were too short, and there was a tendency for the feedback to have a more powerful effect in the younger group. However, estimates of the rate of passing time were not affected by telescoping manipulations, where results were almost identical to those of Wittmann and Lehnhoff, even down to the rating of the current passage of time as slightly fast by both groups.

However, perhaps the kinds of questions being asked in these questionnaires do not capture all of the aspects of time experience that people talk about in casual conversation. It is certainly true that potential changes in time experience with ageing appear to have paradoxical features. My own mother, then aged 88 and living alone, said "The days last forever, but the months flash past." At first sight, this statement is highly paradoxical, as the months which seem to go quickly must be composed of the days which seem to drag, but this can be resolved if the statements about months and days are approached from different judgement perspectives. For example, time locally may seem to drag for an elderly person with a restricted range of activities, but looking back over the month, this very same restricted range of activities leaves the month seeming "empty," with few memorable events that can be recalled, and is thus judged as short. In Chap. 6, I discussed the idea ventured by Wearden, O'Donoghue, Ogden, and Montgomery (2014) that the judgement that "time passes quickly when you're enjoying yourself" was an inference rather than something individuals had actually experienced. My mother's comment suggests, however, that there might be "local" slowing of time in older people, something the questions posed by Wittmann and Lehnhoff and by Friedman and Janssen may not be able to capture, even if it is present.

How can the possibility of changes in local time experience with ageing be investigated? As outlined in an earlier chapter, Droit-Volet and Wearden (2015) decided to probe time experience in the everyday lives of a group of students (mean age 21.75 years) and a group of older people (mean age 72.75 years) using an experience-sampling methodology (ESM). Here, the questions posed in the probe (which is usually a brief questionnaire) are responded to on the basis of immediate experience rather than being based on memory or inference. The commonest current method for conducting ESM is to send alerts via Smartphones or tablet computers, and this was the method used by Droit-Volet and Wearden. Participants received eight alerts each day on a Smartphone between 8:00 in the morning and 8:00 at night. Upon receiving the alert, participants completed a brief questionnaire page with questions asking about the speed of local time, degree of happiness, sadness, arousal, or relaxation, and difficulty of current activity, plus a rating of how much the current activity captured attention, as well as some other questions. Previous questioning had ascertained that both the older and younger group tended to agree to an equal extent with the assertion that time passes more quickly as people aged.

The ESM study probed the degree of happiness, sadness, arousal, relaxation, activity difficulty, and attention capture in the two groups, and there were never any significant differences between the groups on any of these measures—nor, more critically, was there any significant difference between the groups in the judged rate of local passage of time, as assessed by the ESM probes. As such, our main result joins those of Wittmann and Lehnhoff and Friedman and Janssen in finding no evidence of differences in the "speed of time," in our case local time assessed by ESM, with differences in age.

There were some differences, however, between the younger and older groups with regard to relations between the rate of the passage of time and other variables. For example, greater positive affect was positively associated with the passage of time in the younger but not the older group, so being happier made time go faster, but only for the younger people. In contrast, increasing negative affect slowed time in both groups. The effects of arousal were significant for both groups: increasing arousal made time seem to pass more quickly; decreasing it slowed the passage of time. Increasing the difficulty of the activity in which people were engaged increased the pace of time for the younger but not the older group, and likewise, an activity that captured attention made the passage of time seem faster, but only for the student group.

It was unclear why some of the measures produced age differences, in particular why greater positive affect produced faster local speed of time for the younger but not the older group. One possibility is that the difference was related to differences in the activities people performed and which made them happy. Droit-Volet and Wearden also recorded the activities in which individuals were engaged at each alert. In fact, the older individuals were involved in a wider range of activities than the younger group, who spent a good portion of their work week at the university. Although we had information about the activities that each participant was engaged in at the moment of the alert, it was unclear how these very diverse activities could be meaningfully classified, and measures of affect, arousal level, and attentional engagement in the activity were already available. Perhaps some future study can develop a meaningful and valid classification system for the activities performed that could clarify the situations in which age-related differences occurred, although it is not yet clear how this could be accomplished.

Although one should not read too much into a single study, the major finding, that there were no overall differences in passage of time ratings between the older and younger groups, is consistent with previous experiments using the more conventional single-questionnaire probes. ESM, although cumbersome to perform and analyse, seems likely to yield interesting results when comparing other sorts of different groups, or the same groups in different situations. Its use with older people is limited, as it places consider demands on participants' time, and is contingent on their cognitive performance being reasonably intact. Our elderly participants, residents of Clermont-Ferrand and environs living in the community in their own homes, not only completed the probes as diligently as the students, but also engaged in a wide range of activities, as noted earlier, from conventional "retiree" activities such as reading and cooking, to swimming, surfing the Internet, and playing computer games. ESM would be interesting to use on an elderly population with more restricted lives, for example, those living in care homes, but may be difficult to employ for participants with any significant cognitive impairment.

Overall, then, what systematic studies there are yield no evidence whatsoever for changes in time experience with ageing. This conclusion may seem contrary to what "everyone knows," leading to two contradictory conclusions. One is that the phenomenon of "time speeding up with age" is actually a complete illusion, without any evidential reality, so the fact that this is commonly believed proves nothing. For example, the belief that the earth is flat was not only once commonly held, but is compatible with the daily perceptual experience of almost everyone, despite being completely false. An alternative view is that the "speeding up" of time with increasing age is a real phenomenon, and studies to date have simply not posed the appropriate questions to measure it. This begs the question of just what the appropriate questions might be, and in any case, is impossible to disprove: regardless of the number of studies finding no evidence for the speeding up of time with age, they could all be accused of not asking the right question. Perhaps a more interesting and tractable research question is why people believe that time passes more quickly as they age when questioned casually, and why the Lemlich method so reliably generates results consistent with this belief, rather than attempting to prove whether this notion is real or illusory.

Summary

The effects of normal ageing can be observed in the performance of timing tasks. These are largely manifested in the form of increased variability of task performance rather than patterns of performance which differ markedly from those of student-age adults. The increased variability may be an indicator of the slowing of an internal pacemaker with age, but this is far from established. Changes in cognitive functions such as attention and memory with increasing age seem likely to play a role in determining age-related effects on timing tasks, but this issue as yet is little explored. Despite popular belief that "time speeds up as people age," systematic studies with both conventional questionnaires and probes interposed in daily life find no evidence that this is true.

9

Animal Timing

The importance of animal timing in the historical development of modern accounts of time perception cannot be overestimated, even if interest in animal timing per se has been declining for two decades or more. In this chapter, I attempt to trace some of the most interesting and influential research initially carried out in non-human animals, and to look at some ingenious and provocative theoretical models of animal performance on timing tasks, in particular competitors to SET. Animal learning, of which much of the work on animal timing is an offshoot, at one time occupied a central place in academic psychology, but now may be taught in only a cursory manner, if at all, in the undergraduate curriculum of many psychology departments, at least in the UK. This means that contemporary readers may be much less familiar with methods used and ideas derived from Pavlovian and operant conditioning. I cannot provide anything close to an in-depth course in animal learning in a volume on time perception, so a book like Domjan (1993, or later editions) should be consulted for background. I will, however, try to provide the reader with sufficient detail to enable an understanding of the results of the studies I describe. For simplicity of exposition, I have given myself licence to simplify, or even slightly misrepresent, the often complex techniques used in

J. Wearden, *The Psychology of Time Perception*,
DOI 10.1057/978-1-137-40883-9_9

experiments with animals. In particular, some procedures require lengthy pre-training, with gradual approximations to the final method used. I will generally omit any mention of this pre-training, but I will sometimes present simplified versions of the techniques used.

Timing in Early Animal Research: Pavlov and Skinner

One of the earliest manifestations of animal timing in the laboratory was the inhibition of delay identified by Pavlov (1960/1927). Pavlov's methods, usually described as *Pavlovian, respondent,* or *classical conditioning,* involved presenting dogs with pairings of stimuli, one initially neutral, the other a stimulus having some biological significance. For example, the ringing of a bell, sometimes for a minute or more, would be followed by presentation of food, which was independent of any behaviour on the part of the dog. In Pavlovian terminology, the bell was a *conditioned stimulus* (CS), a stimulus which was intended to be neutral in the sense that it did not provoke any particular response (except perhaps an "orienting response" when presented the first few times). The food was an *unconditioned stimulus* (UCS or US), which was intended to provoke some automatic reaction, in this case salivation. The salivation was an *unconditioned response* (UR) to the food, that is, some reliable, naturally occurring response to the US. After a number of pairings, the animal begins to salivate when the bell (CS) is presented, before food is delivered. This salivation, the *conditioned response* (CR), is the principal measure of learning in Pavlovian experiments, and is usually considered to reflect the degree of learned association between the CS and US—the bell and food in this example—with the strength of the CR (in Pavlov's experiments measured in terms of the number of drops of saliva) reflecting the degree of association. The pairing of the CS and US constitutes a Pavlovian trial, and trials are usually separated by inter-trial intervals where no stimuli are presented.

Pavlov's dogs received extensive training, involving hundreds or thousands of trials. When the CS was long, 3 min or more, for example,

Pavlov noted that with increasing training, the salivation became concentrated in the later parts of the CS, with a growing tendency to salivate as the stimulus elapsed, even though it might have occurred earlier in the CS when training was less advanced. Pavlov called this change in salivation over time during the CS *inhibition of delay*. Put in more modern terms, the dog became increasingly sensitive to the length of the stimulus, so its response showed a degree of *temporal control*. Temporal control here means that the response strength, the degree of salivation in this case, depends on some temporal characteristic of the stimulus or experimental situation. In more casual language, the animal's response shows a degree of timing, in that it varies according to the proportion of CS that has elapsed.

Although Pavlov's inhibition of delay was an early and compelling example of animal timing, it has received little attention since, with most modern studies of animal timing using variants of the *operant conditioning* method devised by Skinner (1938).

Typical operant conditioning experiments use rats or pigeons as subjects. The basic apparatus is an operant chamber (or *Skinner box*, a term its inventor disliked), which is equipped with some response device (for rats a lever that they can depress, for pigeons a key they can peck), and some method of presenting small amounts of food. In the case of rats, this is usually a dispenser providing pellets of food; in the case of pigeons, a food hopper which can be raised from time to time, permitting the animal a few seconds of access to grain. The basic premise of operant conditioning is that a lever-press or key-peck can be rewarded by food delivery; so, for example, in a simple case, each lever-press or key-peck is followed by food presentation. In operant terminology, the lever-press or key-peck is a *response* (although there is no suggestion that it is a response to any particular stimulus, unlike in Pavlovian conditioning, so "response" here just means "measurable bit of behaviour"), and the food delivery is the *reward* or *reinforcer*.

Operant conditioning differs in a number of ways from Pavlovian conditioning. Firstly, there are no obvious equivalents of the Pavlovian CS and US. In addition, in many cases, there are no experimenter-controlled trials; the animal is free to respond at any time. Lastly, the guiding principle of operant conditioning, following Thorndike's law of effect, is that the probability of emission of the response is affected by its reinforcement. Put in modern terms, if the response is followed by a reinforcer, it will increase in probability. The typical measure in operant conditioning is response rate, or the number of responses per unit time, and response patterning, or how responses are distributed in time. In some of the studies discussed later, the relevant measure is *choice*: the animal has potential access to two or more responses, such as presses on the left or right lever of the operant chamber, or pecks on keys illuminated with different colours. In these cases, the measure will usually be the proportion of one response over another, such as number of pecks to a red versus a green key.

Traditionally, operant conditioning has not been explained in associative terms. While it might be natural to say that the animal comes to associate the response and reinforcer (and there is evidence that this might be true; see Rescorla, 1991), this explanation of behaviour is rarely used, although associative concepts are used in some theories, such as Machado's *Learning to Time* model (1997), as discussed later.

Animals in operant conditioning experiments, like those used by Pavlov in his experiments, usually receive extensive exposure to the experimental arrangements, often hundreds of daily sessions an hour or more in length. The relation between the response emission and the delivery of the reinforcer is called the *schedule of reinforcement*. There are many such schedules, but from the present perspective, one of the most interesting is the fixed-interval (FI) schedule, mentioned briefly in Chap. 2. On an FI schedule, the first response of the session is reinforced, and this starts a clock. The clock runs for *t* seconds, the FI schedule parameter, and while the clock is running, no responses are reinforced. When the clock times out, the next response that occurs produces food, and the clock restarts.

What is separated in time here, strictly speaking, is the *opportunity* to obtain food after an appropriate response rather than food deliveries themselves, but the pattern of behaviour engendered by FI schedules usually results in food deliveries spaced with close to exact temporal periodicity. The pattern of behaviour found on FI, sometimes after extended training, has been characterized in two ways. One pattern is described as "scalloped." Immediately after food delivery, the animal pauses (the post-reinforcement pause), then responding resumes, initially at a low rate, which accelerates as time elapses in the interval, such that the maximum rate of response occurs just before, or at the moment of, food delivery. The food delivery occasions another post-reinforcement pause, followed by another accelerating pattern of responding, and so on. The second pattern of responding is described as a "break-and-run" pattern, where food delivery results in a pause, followed by a period of high rate responding until the next food delivery, then another pause, and so on. Averaged over a large number of intervals, the pattern of responding shows an increasing rate from near zero immediately after reinforcer delivery, to a peak at or near the time of the next reinforcer. Figure 9.1 shows data from Whitaker, Lowe, and Wearden (2003) for rats responding on FI 30-s and FI 240-s schedules.

Regardless of the pattern of responding proposed, FI schedules clearly engender behaviour which varies in an orderly way with the passage of time after food delivery: the response shows *temporal control* by the time elapsed since food. Furthermore, aspects of the responding vary in orderly ways as the FI schedule parameter varies. For example, the duration of the post-reinforcement pause generally increases as the FI schedule parameter increases, so a rat or a pigeon will pause for a longer time on average if the FI schedule parameter is 60 s than 30 s. Lowe, Harzem, and Spencer (1979), in a particularly careful study, found that the mean pause grew as a power function of the FI value: p, the mean pause, follows a function $p = at^x$, where t is the interval value, and a and x are fitted constants. Because x is usually less than 1.0, the pause is a decreasing fraction of the FI length as the FI value becomes longer. Why this should be true is discussed later.

Variants of the simple FI schedule exist, and these have produced results with considerable theoretical importance in animal timing research. One variant is the two-value *mixed FI* schedule. Here, the time at which the reinforcer becomes available has two randomly alternating and (usually) equally probable values. For example, on mixed FI 30/240-s, the time of reinforcement is either 30 or 240 s after previous food delivery, with nothing at the start of the interval indicating to the animal which value is in force. Performance on this schedule will be discussed later in more detail, but Fig. 9.2 shows some examples of behaviour on a mixed FI 30/240-s schedule from Whitaker et al. (2003).

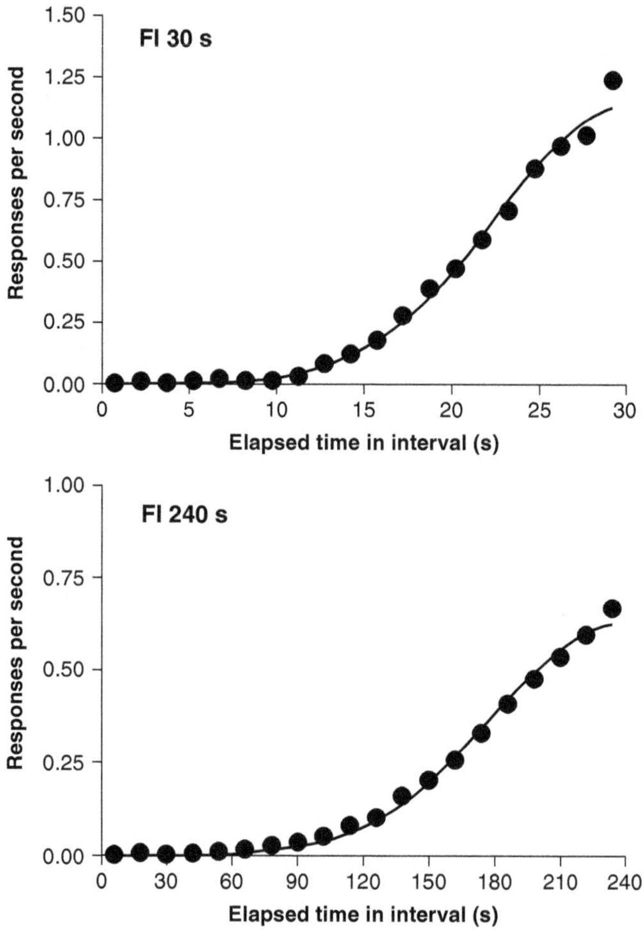

Fig. 9.1 Performance of rats on fixed interval (FI) 30-s and FI 240-s

Another variant is the *peak procedure*. This is a kind of discrete trial FI procedure, where trials begin with presentation of a stimulus such as a light or a tone. On most trials, the start of the stimulus signals a "normal" FI trial: the first response occurring *t* s after stimulus onset is reinforced, the signal is terminated, and this is followed by an inter-trial interval where no stimuli are presented and responses have no consequences. Some trials, however, are "peak trials." Here the stimulus which starts the FI trials

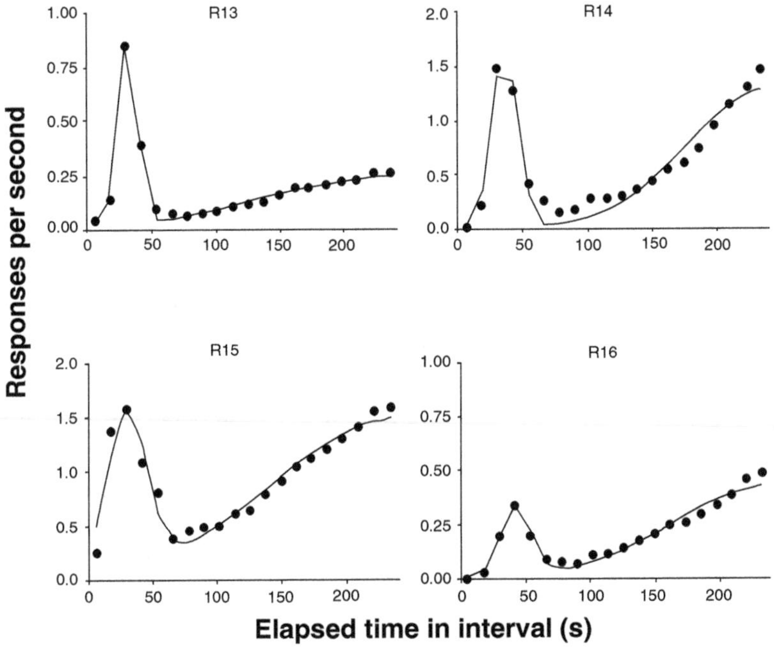

Fig. 9.2 Performance on mixed FI schedules with potential reinforcement times of 30 and 240 s. Data are shown from individual rats tested by Whitaker et al. (2003)

begins, but remains on for longer than the usual FI value, usually two or three times as long. On peak trials, no responses are reinforced, but responses are collected. The typical pattern of behaviour on peak trials is one in which the response rate increases to the level of the usual FI value, then declines to low levels with further increases in elapsed time (Roberts, 1981). Peak trials yield two theoretically important measures. The first is the *peak time*: the time after the start of the trial when the response rate reaches its maximum. The other is the *spread* of the response rate versus the elapsed time function on peak trials. The peak time is usually very close to the time of reinforcement on FI trials, and the spread increases proportionally to the FI value. Peak procedure data thus usually exemplify the two scalar properties of time discussed in Chap. 3: peak time shows mean accuracy, and the change in spread with FI value illustrates the scalar property of variance. Other measures of performance on the peak procedure are also theoretically informative, as will be discussed later.

Skinner developed the FI schedule in his book *The Behavior of Organisms* (Skinner, 1938), and properties of performance on FI schedules were further explored in Ferster and Skinner (1957). However, for largely ideological reasons, discussed in Lejeune, Richelle, and Wearden (2006), he was not concerned with theoretical analyses of temporal control. Another schedule invented by Skinner which has attracted considerable interest in the animal timing literature is the so-called *differential reinforcement of low rate* (DRL). This schedule imposes a requirement for reinforcement on the time between successive responses (inter-response times, IRTs). On DRL *t* s, for example, if the time between responses—the IRT—is greater than *t* s, the second response is reinforced; if not, the second response resets the time requirement. Animals adjust the spacing of their responses under DRL, but long spacing requirements greater than 10 or 15 s pose severe problems, particularly for pigeons when key-pecking is used (see Jasselette, Lejeune, & Wearden, 1990, and discussion later in this chapter).

Operant conditioning techniques not only involve schedules of reinforcement, but can also be used to develop various sorts of discrimination methods. One procedure now commonly used with humans is *bisection*, discussed in earlier chapters. The method used with animals was developed from work by Church and DeLuty (1977). In essence, it is as follows. An operant chamber with two retractable levers is used. Stimuli—for example, tones lasting different lengths of time, or illumination of lights for different periods—are presented, with the levers retracted from the chamber. One stimulus (the *Short* standard) will be 2 s in length, for example, and the other (the *Long* standard) 8 s long. When the stimulus terminates, the levers enter the chamber, and a response on one of them will be reinforced with food after the 2-s stimulus, and a response on the other lever will be reinforced after the 8-s stimulus. Over a few sessions, rats learn this discrimination well, and almost all of their responses are correct and are reinforced. Following this training, a range of stimulus durations is presented, for example, durations from 2 to 8 s in linear or logarithmic steps, and the response to each is noted, with no reinforcers delivered, except sometimes after responses to the 2-s and 8-s stimuli. If we define the response initially trained to be correct after the *Long* standard as the L response, then a psychophysical function can be drawn in terms of the probability of the L response plotted against stimulus duration. This can be analysed to yield measures such as the bisection point and Weber ratio in the manner described for bisection in humans (Chap. 3).

Properties of Animal Timing

If we consider performance on a simple FI schedule, a typical pattern of responding averaged over a number of trials is shown in Fig. 9.1. After reinforcer delivery, no or few responses occur (as this is the period largely occupied by the post-reinforcement pause), and the rate of response then increases until the time of reinforcement. Suppose that the FI value is varied between FI 30 s and FI 60 s. In both cases, after considerable training, the maximum average response rate will occur at or around the time of reinforcement, but the width of the curve will change, being considerably wider in the FI 60-s case than at FI 30 s. The response rate versus time curve can be analysed by simply fitting the left half of a Gaussian function to it, usually with the peak forced to occur at the FI value (Lejeune & Wearden, 1991), and this yields a measure of standard deviation which reflects the width of the curve. Many studies have found that this standard deviation grows proportionally with the FI value, thus reflecting the scalar property of variance. In fact, SET was initially concerned not with human timing but with accounting for data from animal experiments of various sorts, with early exposition found in Gibbon (1977) and Church and Gibbon (1982).

The peak procedure (see Catania, 1970, and Roberts, 1981, for early examples) can also be analysed to test the scalar properties of time. Suppose that, in different conditions, the FI value in the peak procedure experiment varies between 30 and 60 s. Inspection of the response rate versus time functions obtained on peak trials will show that responding increases on average to some peak value, and declines thereafter. Gaussian curves can be fitted to these functions, and these yield two measures; one is the peak location (the time value at which the response rate peaks) and the other is a standard deviation reflecting the width of the curve. In the case described above, the peak location will be close to 30 or 60 s as the underlying FI value changes, illustrating the mean accuracy property of scalar timing. In addition; the standard deviation of the curve will be twice as large at 60 s as at 30 s, so a coefficient of variation statistic (standard deviation/peak location) will be constant as the FI value changes, illustrating the scalar property of variance.

In general, animal timing very often conforms to the two principles of scalar timing, mean accuracy and the scalar property of variance. Lejeune and Wearden (2006) provide a detailed discussion of the evidence, and I will mention here just one of their examples. The first comes from bisection. If the *S/L* standard pairs in bisection are varied, longer time values produce flatter bisection functions, so for example, an *S/L* pair of 2/8 s produces a steeper curve than 4/16 s. However, if the measure of the steepness of the curve is divided by the bisection point for the different cases, the curves are then found to be equal in *relative* steepness, another manifestation of the scalar property of variance. This implies that underlying timing sensitivity remains constant as the intervals used in bisection are varied, provided that the *S/L* ratio is kept the same.

Although reviews of data from animal timing experiments (e.g. Lejeune & Wearden, 2006) find that animal performance conforms to scalar timing in the majority of cases—perhaps the overwhelming majority—early evidence suggested that this was not the case. What looks like non-scalar data is common in the so-called *temporal differentiation schedules* (see review in Platt, 1979). Here, a time requirement for reinforcement is imposed on the responses themselves. I have already mentioned the DRL schedule, where a response must be spaced more than t s from the previous response in order to be reinforced. Another example is *differential reinforcement of response duration*, where a response, such as holding down a lever in an operant chamber, must last some minimum length of time to be reinforced. Many data from temporal differentiation schedules appear to violate scalar timing, in that the average response times occurring, such as the mean spacing between responses or the mean response duration, increases as a power function of the imposed time requirement rather than linearly and accurately with it. So, for example, the mean IRT or mean time produced (m) is often best characterized as $m = kt^x$, where t

is the time requirement and *k* and *x* are fitted constants. The constant *x* is usually less than 1, so the mean response measure becomes a smaller and smaller proportion of *t* as *t* increases. In fact, the situation is even worse. It has been known for many years that on a large number of temporal differentiation schedules, animal performance breaks down when *t* is larger than a few tens of seconds, particularly when pigeon key-pecking is used as the operant response (Kramer & Rilling, 1970). This failure to track imposed time requirements for reinforcement when these are more than 20 seconds or so contrasts markedly with performance on FI, mixed FI, or peak procedure experiments, where animals generally seem to be able to accurately time much longer intervals, sometimes 200 s or more (e.g. Whitaker et al., 2003). If animals like rats and pigeons have the capacity to time intervals hundreds of seconds in length, why do they perform so poorly on temporal differentiation schedules when the time requirements are much smaller?

I proposed a possible explanation in Wearden (1985). Responses like key-pecking and lever-pressing are reinforced with food in temporal differentiation schedules and schedules like FI, so it seems likely that they can sometimes be emitted "at random," that is, without regard to elapsed time in an interval or the time requirement for reinforcement. On a temporal differentiation schedule like DRL, however, these random responses can have severe consequences, as a response which does meet the spacing requirement for the interval resets the time requirement to zero. In my 1985 article, I proposed that animal performance on these schedules is controlled by two processes. One is based on accurate representations of the time requirement for reinforcement, and obeys scalar timing; the other is a certain small probability per second that the response will occur anyway, regardless of time. This latter process "contaminates" measured behaviour, and modelling of this phenomenon has shown that a small probability per second of random responding is sufficient to produce the power function relation between mean response measures and the time requirement often obtained on temporal differentiation schedules, even though underlying timing is scalar.

If the probability of a random response per second is constant, the probability of such a response then increases with time elapsed in the interval, and so, for example, plays a larger role in terminating IRTs when the DRL requirement is long than when it is short. Given a certain probability of random responding, there will actually be a limit to the DRL value that can be reached, as at some long DRL value, all IRTs will be terminated by random responses, and behaviour cannot adjust to longer requirements. This of course is exactly what occurs. For additional discussion of the effect of "random responses" on DRL schedules, see Wearden (1990).

Another situation with power function relations between time values on a schedule and behaviour is the post-reinforcement pause on FI (Lowe et al., 1979), and this can be explained in a similar way. The post-reinforcement pause is terminated by the first response to occur, regardless of whether this is random, so a certain probability of random responding will result in the power function relation between the mean pause and the time requirement, as shown by modelling in Wearden (1985). Measures like the peak location on the peak procedure are not systematically affected by these "stray" random responses, so they can generally track time requirements accurately, even at long values, and this is also true for the typical average response rate versus elapsed time measure on FI, which peaks at the time of reinforcement even when this is very long (Lejeune & Wearden, 1991).

The "random" responses proposed by this model are hypothetical. Presumably, an experimenter would be unable, by observation, to distinguish a key-peck that occurred at random from one that was "deliberate" and governed by a decision process such as that proposed by SET. How, then, can this idea be tested?

Jasselette et al. (1990) reasoned that some responses, because of their ease of emission, were more likely than others to be emitted "at random." A key-peck or lever-press requires little in terms of energy, and hundreds may occur in an interval such as FI with a long time to reinforcement. Suppose, however, that the response was much more difficult to make. Now, the response would be less likely to be emitted at random, and the underlying (presumably scalar) nature of the timing would be observed. The authors tested this by training pigeons to stay on a perch for a period of time (in a temporal differentiation of response duration condition) or to produce two perching responses with a certain length of time between them (in a DRL condition). This was a laborious response for the pigeons, and could only be emitted at a low rate. When this perching response was made, the animals could track timing requirements much longer than normal for temporal differentiation schedules when key-pecking was used (Jasselette et al., 1990, employed values up to 70 s). Furthermore, response distributions were scalar, in that the mean time produced was generally accurate, with a constant coefficient of variation. For similar results, see Lejeune and Jasselette (1986).

Overall, then, performance on temporal differentiation schedules may not violate the principles of scalar timing of behaviour as strongly as it initially appeared to, substantiating the claim that scalar timing is a good characterization of animal timing in many situations.

Explanations of Animal Timing According to SET

As mentioned in an earlier chapter, SET, which has enjoyed considerable success since the early 1990s as an account of human performance on timing tasks, was initially invented to explain data from animal timing studies. Some of these data came from standard reinforcement schedules like FI and its variants such as the peak procedure (Roberts, 1981). Others came from avoidance tasks involving timed responses (see Gibbon, 1979), and yet others from tasks developed to provide a kind of animal psychophysics of time, such as bisection (Church & DeLuty, 1977) and temporal generalization (Church & Gibbon, 1982).

Although very seldom used with animals, the task of *temporal generalization* developed by Church and Gibbon (1982) not only illustrates the basic properties of scalar timing in animals, but also provides a basis for understanding some of the more complex applications of SET to animal timing that will be discussed later. In the original article, the subjects were rats, and the apparatus an operant chamber with a single retractable lever. The chamber was illuminated by a single light, and switching this light off for various periods of time constituted the stimulus to be timed. Training proceeded in several stages, which I will not discuss here, but the final procedure of the illustrative Experiment 1 was as follows. The standard or target stimulus was 4 s of darkness. When the light came on again after the dark period, the lever entered the operant chamber, and a single press was reinforced with food. If the stimulus was longer or shorter than 4 s (values varied between 0.8 and 7.2 s or 1.0 and 16.0 s in Experiment 1, where the effects of linear or logarithmic spacing of non-standard stimuli were used), the lever-press was not reinforced. In Experiment 1, the probability of responding peaked at the standard, 4 s, and declined symmetrically as durations deviated from the standard in the linear spacing condition. In other words, a 4.8-s stimulus produced the same probability as a 3.2-s stimulus, a 5.6-s stimulus the same as 2.4 s, and so on. However, there was substantial response probability (around 0.25) at stimuli remote from the standard, something confirmed in another experiment where the standard was 4 s but the longest non-standard duration used was 32 s.

Other experiments varied the standard duration (e.g. over values of 2, 4, and 8 s) and varied the probability of reinforcement for responses after the standard (which decreased overall response probability), as well as other conditions not discussed here. One feature present in the data was superimposition: when response probability versus stimulus duration functions were expressed on the same relative scale (with all durations expressed as a fraction of the standard), data from conditions with different standard values were well superimposed (Church & Gibbon, 1982, Fig. 10, p. 183), illustrating good conformity to the principles of scalar timing.

The theoretical analysis of temporal generalization performance provided principles which were later used to account for other behaviours in SET-consistent models. A fundamental notion here is that the animal had a representation of the standard duration, s, which was a Gaussian distribution with mean s, but some coefficient of variation (standard deviation/mean), c. This results from variability in the storage of the standard, s, either because (a) the standard is timed correctly but has added variability during the storage process (e.g. s is multiplied by a Gaussian value with a mean of 1.0 and coefficient of variation, c), or (b) scalar variation in pacemaker speed from one interval to another causes the effective representation of the standard to differ from trial to trial. Classical SET (e.g. Gibbon, Church, & Meck, 1984) preferred the former explanation.

A random sample, s^*, from this memory distribution is selected on each trial, and this is compared with the representation of just-presented stimulus duration, t, assumed to be timed without error. A response occurs if

$$\text{abs}\left(s^* - t\right) / s^* < b^* \qquad (9.1)$$

where b^* is a random sample from a threshold with mean, b, and some trial-by-trial variance, and *abs* indicates absolute value (i.e. difference regardless of sign). Here, the left side of Eq. (9.1) gives a measure of the difference between s and t, and when this value is less than b^*, the standard and comparison durations are "close enough" to warrant a response.

Inspection of Eq. (9.1) reveals some of its properties. Ignoring the trial-by-trial variability of s and b, the smallest absolute difference between s and t—and thus the value most likely to be below b and therefore provoke a response—is when $s = t$, so responding should be maximal when the comparison duration equals the standard, and it is. Dividing the absolute difference $(s - t)$ by s results in superimposition. If the standard is doubled, for example, then the absolute difference between s and t must be doubled to produce the same difference measure. The equation also predicts symmetrical gradients: the same difference between s and t will provide the same difference measure regardless of whether s is longer or shorter than t. In addition to Eq. (9.1) (their terminology was slightly different, but the equation is the same), in order to model the data, Church and Gibbon (1982) also added two "attention" parameters, one involving the probability of attending to the signal on the trial, and the other the probability of responding in the absence of attention to the signal. These parameters were needed to model the high response probability sometimes observed at durations remote from the standard. A very similar idea—in fact, actually directly borrowed from Church and Gibbon (1982)—was used by Droit-Volet, Clément, and Wearden (2001) to model temporal generalization performance in children, as discussed in Chap. 7.

The temporal generalization task thus illustrates all aspects of the cognitive architecture of SET as applied to animal timing. Stimulus durations are timed by the internal clock, the two memory stores are used (reference memory for representations of the standard, working memory for the comparison), and a decision process [given in Eq. (9.1)] generates responding. Superimposition is generated by the "normalization" (i.e. division) of the absolute $s - t$ difference by the standard s.

The principles governing SET's treatment of performance on FI are very similar. On a simple FI schedule, the first operant response occurring at more than *t* s from the previous food delivery is itself reinforced, and this reinforcer delivery restarts the interval. For simplicity, let us consider an FI 60-s schedule. SET proposes that the internal clock is reset to zero with reinforcer delivery, so the clock times up from zero, and pulses accumulate in the working memory. At a time just longer than 60 s, the reinforcer is delivered, so the elapsed time on that trial is entered into working memory as a representation of the time of reinforcement. As in the case of temporal generalization, the representation of the standard is then also entered into reference memory with some trial-by-trial variation (resulting either from a transformation when the value is stored or variable pacemaker speed from trial to trial, as in the temporal generalization case), but again, a distribution with a mean equal to the FI value and a coefficient of variation of *c* is built up in reference memory. Elapsed time resides in a constantly changing working memory/accumulator module (although working memory and the accumulator were distinct in Gibbon et al. (1984), more recent versions of SET have combined them into a single module), and readings from this module are compared with a sample taken from reference memory. The model predicts that responding will begin when the value in working memory and the value drawn from reference memory are "close enough," with the decision rule the same as for temporal generalization [i.e. Eq. (9.1)].

Performance on the peak procedure is accounted for in a similar way. During the peak trial, an elapsed time occurs which is sufficiently close to the stored time of reinforcement (learned on the FI trials), responding then begins, and continues until the elapsed time ceases to be "close enough" to the time of reinforcement, at which point responding ceases. It may seem from this crude outline of SET's account of peak procedure performance that a SET-based model can, at least in principle, account for the elapsed time at which the animal commences responding on peak trials (the start time), and also for the time at which responding ceases (the stop time), and therefore will also predict the total length of time during which the animal responds (the spread) and the midpoint of the response period (the middle). In fact, considerable effort has been expended by some SET theorists on the details of responding on the peak procedure, starting with an article by Gibbon and Church (1990). The arguments used to predict start/stop patterns of behaviour on the peak procedure are quite complex, even in a simplified form, and I discuss them in outline form in the appendix. For further details, the reader is referred to Gibbon and Church (1990) and Church, Meck, & Gibbon (1994).

After the complexities of predicting start and stop times on a peak procedure task, SET's treatment of a two-value mixed FI schedule might at first seem simple. Suppose we have two equally probable values FI A and FI B (where B > A). Whitaker et al. (2003) showed that timing on such schedules was generally scalar, the peaks of responding occurred at the times of reinforcement, and the width of the response rate versus time functions centred on times A and B growing in proportion to the value of A and B. Some data are shown in Fig. 9.2. All this seems consistent with SET's treatment of FI and the peak procedure. At some time near A, the difference between the current elapsed time in the trial and the memory of A will be "close enough," according to Eq. (9.1), so responding will start, and will cease (as on the peak procedure) at some later time when the elapsed interval is no longer "close enough." Later in the interval, elapsed time will then be "close enough" to B, so responding will start again. Overall, then, responding should generally increase up to time A, then decline, only to increase again until time B, with the widths of the response versus time functions having scalar properties. This is what actually happens in practice (Whitaker et al., 2003; see Fig. 9.2), so SET's treatment appears unproblematic.

However, Machado, Malheiro, and Erlhagen (2009) point out a potential logical difficulty in SET's treatment of mixed FI. In order to explain the response pattern actually observed in terms of SET theorizing, some way has to be found to direct the comparison of the elapsed time first to the memory of FI A, then later to the memory of FI B. Leak and Gibbon (1995) proposed separate memory stores for the memories of times A and B. Initially, the animal is assumed to use the memory of A, then later the memory of B. But, as Machado et al. (2009) note, it is not clear how this is possible, as there are no separate external cues associated with A and B, just the "passage of time." Further discussion of this issue and others related to SET's treatment of the timing of multiple intervals can be found in Machado et al. (2009).

SET's treatment of bisection, in contrast to its treatment of mixed FI, seems free from Machado et al.'s "logical problem," as in the case of bisection, there are two distinct responses, so counts can be directed appropriately to the two different memory stores, one associated with each response. Suppose that the standard durations are 1 and 4 s, with lever presses on the left and right levers of an operant chamber reinforced after 1 or 4 s, respectively. If we identify the left and right responses as SHORT and LONG responses, then after the 1-s stimulus, the counts produce some memory distribution centred around 1 s, and this is associated with the SHORT response, while similarly, a memory distribution centred on 4 s is associated with a LONG response. When intermediate stimulus durations are presented during bisection testing, a decision rule is needed to determine whether a SHORT or LONG response should be emitted. Gibbon (1981) gives a comprehensive mathematical account of potential bisection decision rules, to which the interested reader is referred, but the decision rule generally employed in SET's treatment of bisection in animals is to respond SHORT if ($t/S < L/t$) and LONG if ($t/S > L/t$), where t is the current duration, and S and L samples from memories of the short and long standards. The point at which 50 % of LONG and SHORT responses should occur is when ($t/S = L/t$) or when $t = \sqrt{(S \times L)}$—in other words, at the geometric mean of S and L. Geometric mean bisection is the result normally found when animals are tested on bisection tasks (see Church & DeLuty, 1977, and Meck, 1983, for demonstrations).

Overall, therefore, SET provides a structure that can account for performance on a range of timing tasks, only the most common of which have been illustrated here. Some problems, particularly concerned with the assignment of counts to different memory stores, can be identified, but in general, models based on SET principles can be devised which will fit data well. Of course, this is perhaps to be expected, because the ability to change aspects of the model to fit the task—for example, changing decision processes—provides SET with a good deal of flexibility, as some have complained (e.g. Staddon & Higa, 1999). Nevertheless, there are consistencies between treatments of similar but different tasks, as illustrated by the use of the same decision equation in determining whether two times are "close enough" in tasks such as temporal generalization, FI, and peak procedure.

Despite its success in fitting data, SET is not without its competitors, and the next section discusses the best-developed of these, the *Behavioural Theory of Timing* and the *Learning to Time* model.

Competitors of SET

The Behavioural Theory of Timing

Radical behaviourists of the Skinnerian sort are reluctant to account for behaviour in terms of internal processes such as clock operations and decisions in contrast to a cognitive theory like SET. Lejeune et al. (2006) review some of the history of attempts by those sympathetic to radical behaviourism to come to grips with the theoretical problems that time seems to pose for them. For example, suppose in an experiment with pigeons, responses are reinforced if the key is illuminated with a green light, but no reinforcers are delivered for responses if the key is red. Pigeons will quickly learn this, and will respond only if the key is green. This discrimination poses no explanatory problem for radical behaviourists, as external stimuli, red and green, signal different reinforcement contingencies (presence or absence of food after responses), so behaviour is clearly under the control of external environmental stimuli, which "set the occasion" for responding. However, when we look at a well-trained animal on a simple FI schedule, the animal will respond scarcely at all for some period after reinforcer delivery, and yet later in the interval will respond rapidly. The problem here is that there is no external stimulus enabling the animal to distinguish early parts of the interval from later parts: the stimulus seems to be "internal," some inner process sensitive to elapsed time in the interval, SET's internal clock, for example. Traditionally, explanations of behaviour in terms of internal processes are forbidden in radical behaviourism, so some means must be found to account for temporally regulated behaviour in the absence of any changing external stimuli, if "time itself" cannot be admitted as a stimulus.

Killeen and Fetterman (1988) grasped this theoretical nettle in their *Behavioural Theory of Timing* (BeT). The starting point of this theory was an experiment by Skinner (1948), who delivered food to hungry pigeons every 12 s, without the need for any particular response from them. After some hours of exposure to this condition, all pigeons were observed to have adopted patterns of stereotyped behaviour in the period between food deliveries: one might turn in circles, for example, while another might show stereotyped movement patterns such as regular pacing of the chamber. Skinner called these behaviours "superstitious." The pigeon must be doing something when food is delivered. Whatever the response is, it will be increased in probability by food delivery, and is thus more likely to occur when the next food delivery arrives, further increasing the response probability. The blind operation of the law of effect thus "accidentally" strengthens some behaviours or patterns of behaviour, even though these actually play no role in the delivery of the reinforcer, which would occur regardless of the behaviour produced. Skinner drew analogies between these behaviours and human superstitions: some actually non-causal behaviour is produced, and by chance is followed by a desirable event, which strengthens that behaviour. This causes it to be repeated, thus more likely to have been emitted when another desirable event occurs, and superstitions therefore develop even though the behaviour actually has no relation to the event it is supposed to produce.

Skinner proposed that any behaviour could be "accidentally" reinforced in this way, although later work (Staddon & Ayres, 1975; Staddon & Simmelhag, 1971) qualified this somewhat. Stereotyped behaviour sequences certainly developed when animals received periodic food independent of their responses, but there seemed to be constraints on just which behaviours would occur at different times in the sequence. In particular, behaviours associated with consummatory activities, like pecking in pigeons or approaching the food hopper in rats, tended to cluster towards the end of the sequence, before food was delivered, whereas other activities, such as stereotyped movement patterns, tended to occur earlier.

Killeen and Fetterman's BeT assumed that the kinds of behaviours that Skinner and others observed when food was periodically delivered, which came to be called *adjunctive* behaviours, played a critical role in the genesis and maintenance of temporal control. Consider a simple FI schedule. BeT's explanation of behaviour was that a sequence of unobserved adjunctive behaviours occurred, each of which was a cue for the next, so the animal would emit behaviours B_1, B_2, B_3...B_N, with one of these being the cue that led to a period of key-pecking or lever-pressing, which would then be followed by reinforcer delivery. The reinforcer delivery at the end of the chain would maintain the behaviour sequence by means of the law of effect.

According to BeT, then, an animal on a schedule like FI is not really "timing" the interval at all, in the sense that it has no direct measure of elapsed time per se, such as the readings of the internal clock in the SET model; it possesses merely a sequence of adjunctive behaviours, one of which serves as a discriminative stimulus for emission of the reinforced response. Not only is there nothing measuring time elapsed in the interval, but there is also no equivalent of SET's memory or decision processes. The "memory" of the time of reinforcement is embodied in the adjunctive sequence, and the animal does not "decide" to emit the operant response at the end of the interval; the observed response is merely triggered by one of the adjunctive behaviours preceding it.

BeT thus maintains a "behavioural" character consistent with the ideology of radical behaviourism, and dispenses with SET's elaborate internal clockwork of clock, memory, and decision processes. However, it does not entirely dispense with internal processes, as in fact a pacemaker is proposed, the successive "ticks" of which move the animal from one adjunctive behaviour to the next. It is important to distinguish this pacemaker from that of SET's internal clock: the BeT pacemaker is not "read": its ticks, which are supposed to be indistinguishable from one another, are not accumulated, and do not provide any "measure" of elapsed time.

A reader coming across this theory for the first time might assume that it is easily tested by observation. If we watch rats or pigeons on an FI schedule, for example, we should then clearly see the adjunctive sequences that BeT proposes. In fact, the situation is more complicated than that, and I will leave this issue until later in this section. BeT's pacemaker is of the Poisson type, that is, ticks are emitted at random, with some averagely constant rate. This poses an immediate problem for BeT. If the adjunctive sequences are driven by a Poisson pacemaker, then a greater number of adjunctive behaviours will occur at longer intervals than shorter ones. This will give the resulting behaviour the character of Poisson timing: a relative measure of variability such as the coefficient of variation will be smaller at longer intervals than shorter ones. This violates the scalar property of variance which, as we have seen, is very often obeyed by animal timing (Lejeune & Wearden, 2006), so some means of reconciling the principles of BeT with scalar timing must be devised. Killeen and Fetterman proposed a simple solution: the rate of the pacemaker is proportional to the rate of reinforcement delivered by the experimental situation. In the case of FI, the longer the interval, the lower the rate of reinforcement delivered, so the pacemaker, for example, should run more slowly on FI 60 s than FI 30 s. In its simplest form, BeT proposes that the *number* of adjunctive behaviours is constant regardless of the interval length: the behaviours each last longer and the adjunctive sequence is traversed more slowly if the rate of reinforcement is lower or the FI longer. This mathematical adjustment of pacemaker rate enables BeT to predict behaviour which shows the scalar property of variance. Figure 9.3 shows calculated pacemaker rates from Killeen and Fetterman (1988) as a function of rate of reinforcement.

It is important to note that the results shown in Fig. 9.3 come from the *calculated* pacemaker rates needed to model the data; the pacemaker rate is not measured in any direct way. Direct observations of adjunctive

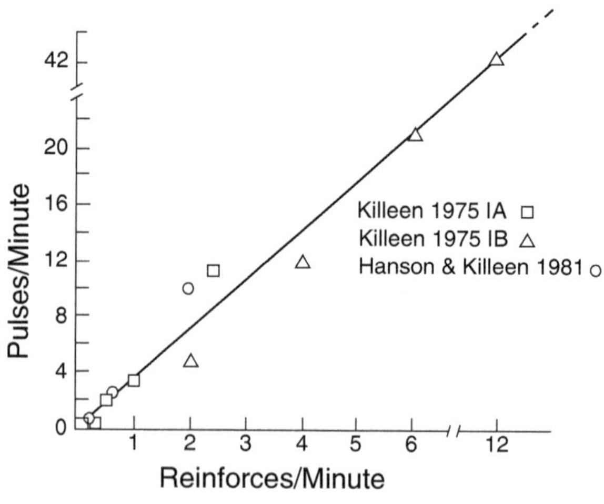

Fig. 9.3 Inferred pacemaker rates (pulses per minute) plotted against reinforcement rate (reinforcers per minute), from Killeen and Fetterman (1988), Fig. 3

behaviour, which might yield information about the actual rate of the proposed pacemaker, have been rare and are discussed later.

Research surrounding BeT has generally focussed on three areas. One is showing that the ideas embodied in BeT, despite their apparent simplicity, can be used to model behaviour on a range of timing tasks. I have used FI as an illustration here for simplicity, but models of other tasks, such as bisection, can be readily generated (see Killeen & Fetterman, 1988, for example). The second focus of interest has been in obtaining evidence that the pacemaker rate changes, as the theory requires, when reinforcement rate is changed.

A study by Fetterman and Killeen (1995) provides what is perhaps the best evidence that the pacemaker rate shifts with the reinforcement rate in a manner similar to that required by BeT. The authors used a "categorical timing" procedure, which in simplified form operated as follows. Pigeons worked in an operant chamber with three response keys, which varied in terms of when reinforcement was available for responses on them from the beginning of the trial, where the beginning was signalled by the illumination of one of the keys. The right, centre, and left keys were associated with a short, intermediate, and long interval, respectively, but only one reinforcer was available on each trial. For example, in one condition (base 4), the keys "paid off," if reinforcement had been assigned to them on that trial, after 4, 8, or 16 s, respectively. Other conditions were similar except for the time values used (e.g. 8, 16, and 32 s, base 8).

Average response rates on the different keys varied according to time elapsed in the interval. For example, in the base 8 conditions, pigeons would initially show a low response rate on the short key, and this would then increase to a peak around 8 s, only to decline and be replaced by increased responding on the middle key, which would peak around 16 s and then decline, itself replaced by responding on the long key, which generally continued until reinforcer delivery. Fetterman and Killeen's article involves elaborate modelling of many features of responding, including times of transition from one key to another, but for purposes here, I will describe only one manipulation.

This involved changing the probability of food delivery. In most conditions, the probability was 1.0, that is, one of the keys would "pay off" on the trial. However, some conditions involved a reduction in the reinforcement probability to 0.25, meaning that a reinforcer was delivered in only one in four trials. This, of course, changed the reinforcement rate for the condition, and so, according to BeT, should have changed the pacemaker rate. The experimental design enabled observation of behaviour after shifts from high- to low-density reinforcement (1.0–0.25) and vice versa (0.25–1.0). The data presented as an illustration in Fetterman and Killeen's article come from responses on the middle key of the base 4 and base 8 schedules. When reinforcement probability was reduced from 1.0 to 0.25, the peak of the response rate versus time function was shifted to the right: that is, the animals' responding peaked later under a 0.25 condition than a 1.0 condition. When the transition was in the other direction, the opposite shift occurred; when transferred from a 0.25 to 1.0 probability, the animals' response peaked earlier than before.

These changes in peak location with shifts in reinforcement density are consistent with BeT's predictions, at least qualitatively. When the reinforcement density decreases, the pacemaker slows, and the adjunctive sequence takes a longer time to traverse, resulting in later peak responding. When the reinforcement density increases, the pacemaker runs faster, so the animals' responses peak earlier. However, the shift in peak responding was much smaller proportionally than the change in reinforcement rate. The rate varied fourfold (0.25–1.0), but the peak shifts were narrower, on the order of 10 or 20 %. Fetterman and Killeen (pp. 60–61) acknowledge this difficulty and suggest possible ways to circumvent it.

Two-value mixed FI schedules, where a response is reinforced according to either FI A or FI B (with B > A), with the values changing unpredictably, can also be used to test the proposition that pacemaker rate varies with reinforcement rate. Consider a schedule like FI 30-s and FI 240-s, used by Whitaker et al. (2003). Here, a reinforcer is available either 30 s or 240 s after the previous reinforcer delivery. The two components of the schedule (30 s and 240 s) obviously deliver very different rates of reinforcement (120 reinforcers per hour for FI 30-s but only 15 per hour for FI 240-s). If the pacemaker rate is constant, the 240-s component should be timed more precisely than the 30-s one. A problem for the animal is that it does not know which schedule is in force immediately after reinforcement, so presumably it uses a pacemaker with a constant rate, perhaps one based on the average rate of reinforcement available on the mixed FI schedule. Whitaker et al. (2003) looked at the spread of the response versus time function for the 30-s and 240-s components of the mixed FI (as well as other values in different experiments), and found little deviation from scalar timing: that is, the timing of the longer components was no more precise than that of the shorter components (see Fig. 9.2, for example). This is difficult to reconcile with a constant-rate pacemaker of the type proposed by BeT, so some means of changing the pacemaker rate in the different components is needed in order to produce the scalar timing found in data. As mentioned above, it is unclear how this can be accomplished, as the animal is incapable of predicting which component of the mixed FI is in force. One suggestion which might reconcile the data with BeT is that the animal changes the pacemaker rate during the interval. For example, the animal starts the interval with a pacemaker corresponding to FI 30-s (or whatever the shorter component is), and then at some point in the interval decides that the 240-s component is in force, and slows the rate of the pacemaker. There is little evidence for this, however, although the drop in the response rate at around 60 s on the 240-s interval of mixed FI 30/240-s (see Fig. 9.2) suggests that the animal "knows" that it is not going to be reinforced at the time shortly after 30 s, and so could adjust the pacemaker rate accordingly.

The final focus of interest for BeT, but one which has received much less research attention than someone encountering this theory for the first time might expect, is actual observation of the behaviours occurring in conditions where temporally regulated operant responses like key-pecking were measured. That is, do the adjunctive sequences proposed by BeT actually exist, and do they play the mediating role that the theory proposes?

Haight and Killeen (1991) provide perhaps the clearest evidence generally compatible with the premise of BeT. Briefly, pigeons were exposed to multiple fixed-time (FT) schedules. A multiple schedule is one in which the schedules of reinforcement operating change from time to time, but with the change being signalled—in the case of Haight and Killeen, by a change in the key colour in an operant chamber between red and green. On an FT schedule, the reinforcer is delivered at a fixed time after the previous reinforcer delivery, with no response necessary to produce it. The baseline condition in Haight and Killeen's study was multiple (mult) FT 15-s/FT 15-s, that is, in both components of the schedule (signalled by red and green key-lights), food was delivered every 15 s, with no response requirement. Like the pigeons in Skinner (1948), Haight and Killeen's subjects developed sequences of superstitious adjunctive behaviours between reinforcer deliveries. When these had become established after 75 sessions of training, the schedule associated with one of the components was changed to either FT 5-s or FT 45-s, with the other component remaining unchanged at FT 15-s. The logic of this manipulation was that changing the schedule to FT 5- would *increase* the rate of reinforcement overall, above what it had been on mult FT 15-s/FT 15-s, whereas changing the schedule to FT 45-s would *decrease* it. Increasing the reinforcement rate should increase the pacemaker rate, so the adjunctive sequence established during the baseline (mult FT 15-s/FT 15-s) should proceed more rapidly, and decreasing the reinforcement rate should have the opposite effect. To test this idea, the adjunctive behaviours occurring during the unchanged FT 15-s component were observed. In general, the sequences behaved as predicted: traversing more rapidly when the reinforcer rate increased and more slowly when it decreased (although Haight and Killeen do not present any quantitative data supporting this claim).

Perhaps the most elaborate observational study of adjunctive behaviours during animal timing comes from Lejeune, Cornet, Ferreira, and Wearden (1998). The subjects in this study were gerbils, and the task was differentiation of response duration, where the response of interest was residence on a platform. To obtain the reinforcer, the animal had to mount a platform, stay on it for at least x s, and then dismount. The time requirement for reinforcement was varied over values from 2 to 15 s. In addition to imposing this temporal differentiation contingency on platform residence, all behaviours emitted by the animals were observed and classified into 12 categories according to a coding scheme. In general, the time spent on the platform increased linearly and accurately as the time requirement for reinforcement increased, and the coefficient of variation of the time spent was constant except at the lowest value (2 s), indicating good conformity of platform residence time to the two scalar properties of time, mean accuracy, and constant coefficient of variation.

The number of adjunctive behaviours increased as the time requirement increased, but the increase was far from proportional. The time requirement changed from 2 to 15 s, a 7.5-fold change, but for different animals the number of adjunctive behaviours doubled or tripled, and for one animal remained about the same. The rate of adjunctive behaviours increased with increasing reinforcement rate, so adjunctive behaviours were emitted more rapidly when the animal was reinforced more often, as in Haight and Killeen's (1991) study. The latter result fits well with BeT's predictions, and may provide direct observational evidence of the kind of changes in pacemaker rate in different conditions that BeT proposes. The former result is not completely consistent with BeT, which would require the number of adjunctive behaviours to remain constant as the interval timed changed, but the increase in the number of adjunctive behaviours was far from proportional to the interval length, so the result is still partially consistent with BeT.

Results which were less in accord with BeT involved observation of the existence of "repeats," or adjunctive sequences where one of the behaviours was repeated, usually after another behaviour had occurred between the repeated behaviours. As mentioned above, according to BeT, the adjunctive behaviours act as cues for other adjunctive behaviours, and eventually the reinforced response, and the pacemaker "ticks" are not accumulated, do not directly measure elapsed time, and are supposed to be identical, so they give the animal no temporal information whatsoever. The existence of repeats renders the adjunctive behaviours very ambiguous cues, but one possibility, suggested by Lejeune et al. (1998), is that the repeated adjunctive behaviours, while they may appear identical to a human observer, are somehow distinguished by the animal, and thus can serve as the distinct cues needed by BeT. The authors also found that adjunctive sequences varied from one trial to another, even when the measured response was accurately timed. Strictly speaking, BeT does not require the adjunctive sequences to be the same on all trials. It would be possible to have a number of adjunctive sequences, all of which resulted in the measured response occurring at a certain time, just as a person can travel from X to Y taking different routes. However, if the adjunctive sequences are too variable, the argument that they have been strengthened by superstitious reinforcement becomes difficult to sustain, and their causal status is called into question.

In general, the results obtained by Lejeune et al. (1998) offered some support for BeT, although they illustrate the difficulties of testing BeT by observation. Other problems with examining the relation between adjunctive behaviours and measured operant responses are logical in nature. For example, suppose that adjunctive behaviours actually play no role at all in controlling measured responses, which are instead triggered by an internal clock of the SET type. Even though the adjunctive behaviours are not "necessary" here, the animal must be doing something between food deliveries, and it is possible that the delivery of reinforcement strengthens certain behaviours (as Skinner, 1948, proposed), or even organizes the adjunctive behaviours into regular chains. How can it be conclusively demonstrated that the adjunctive behaviours play a causal role? This was indeed the question raised in an early review of adjunctive behaviours by Richelle and Lejeune (1980), who concluded that adjunctive behaviours occurred, but may not play a causal role: for example, if they were prevented or disturbed, measured operant responses showed only small temporary changes.

A more recent report by Gouvêa, Monteiro, Soares, Atallah, and Paton (2014) used modern video technology to examine the head trajectories of three rats performing a kind of temporal categorization task. Unfilled auditory stimuli ranging in length from 0.6 to 2.4 s were presented; if the stimulus was longer than 1.5 s, the animals were reinforced for poking their noses into a left hole in the operant chamber, and for shorter durations they were reinforced for a right response. Nose-poking behaviour showed a typical bisection-like ogival increase in left responses as stimulus duration increased. The authors also used a kind of motion-capture technology to observe the animals' head location trajectories over time. Head location patterns were idiosyncratic between rats but reproducible within each subject, and were shown to predict choice responses, even for stimuli of the same duration. Although this study shows strong relations between choices and movement patterns, it is unfortunate that the response measured was nose-poking and that only head movements were monitored. It is perhaps unsurprising that head movement patterns for a rat about to poke its nose into a hole in either the left or right side of a panel in an operant chamber would be different before the response was actually measured. Nevertheless, the results offer partial support for BeT, although, as the authors themselves point out, they do not rule out the possibility that some central cognitive timer controlled response choices rather than the movements themselves.

Notwithstanding the surprising difficulty in conclusive testing this alleged "behavioural" theory by actual observation of behaviour, BeT in general is a particularly ingenious attempt to account for many of the results obtained from animal timing experiments, with a model much less cognitive than SET and which dispenses with much of the "clockwork" of SET, such as memory and decision processes. One clear limitation of BeT is that it cannot be easily applied to timing in humans, particularly when brief durations are timed. When individuals carry out temporal generalization or bisection tasks where all timed stimuli are less than 1 s in length, it seems highly unlikely that some sequence of adjunctive behaviours has time to occur. An additional problem is that humans are able to discriminate stimulus durations—provided they are sufficiently different (e.g. 100 and 900 ms) —the first time they encounter them, so there can be no prior history of reinforced adjunctive behaviours, even if these could be imagined. BeT is an example of a theory involving "timing by learning," in this case learning to develop an adjunctive sequence as a result of "superstitious" reinforcement. This learning process must take some time, so it is difficult to explain the discrimination of stimulus durations the first time the stimuli are presented. This difficulty is not so evident in animal timing, as the animal will receive extensive training, so the development of adjunctive sequences remains a possibility, but is an important issue for any "timing by learning" model (e.g. that of Matell & Meck, 2004) when it is applied to humans.

Learning to Time

Armando Machado's (1997) *Learning to Time* (LeT) model is sometimes considered an offshoot of BeT, although it is clearly less "behavioural" in nature. LeT has the unique feature of predicting not only *when* animals will respond, but *how much* they will respond as well. In other words, it provides a model of response rate as well as temporal control, whereas SET and BET predict when an animal will respond but not the rate of response which will occur. The ability to predict response rate is both a strength and weakness of LeT. Its ability to predict response rate is impressive, but the predictions are not always correct, as we will see later.

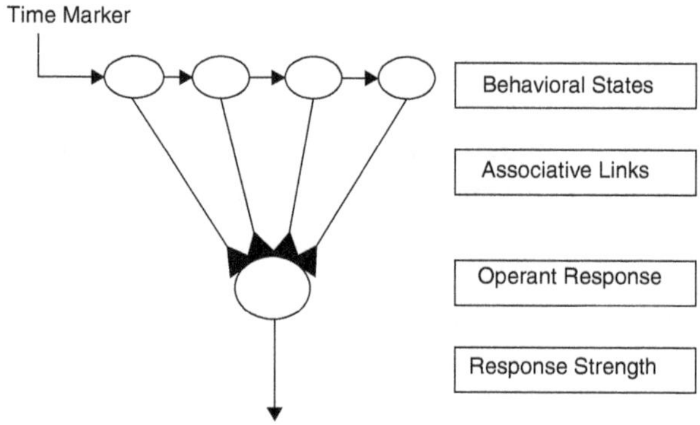

Fig. 9.4 Basic structure of LeT. A time marker initiates a series of behavioural states, each having an associative link—the strength of which can be modified—with the measured operant response. The total strength of the associative links from the states activated at any given time determines the response rate

The ability of LeT to predict response rate derives from its basic structure, illustrated in Fig. 9.4, taken from Machado (1997). A time marker such as reinforcer delivery initiates a series of states which rise and fall in activation level (as shown in Fig. 9.5, also taken from Machado, 1997). Each state has an associative link with the measured operant response, although this link may have zero strength, meaning that the operant response is not emitted when that state is active. The response rate at any given time is determined by the sum of the strengths of the associative links between the states active at that time and the measured operant response.

Like BeT, LeT is perhaps best illustrated initially with respect to performance on a simple FI schedule. The reinforcer delivery which starts the interval triggers the unfolding of successive, partially overlapping "states" (Fig. 9.5). Each of these states rises in activation over time from near zero, then falls again to near zero, and the duration of the activation of the state is dependent on when it is activated, so states activated later in the interval last longer; for example, the state maximally activated 10 s after reinforcement is active for a longer time than the state maximally activated at 5 s, and so on. In fact, the length of time during which the

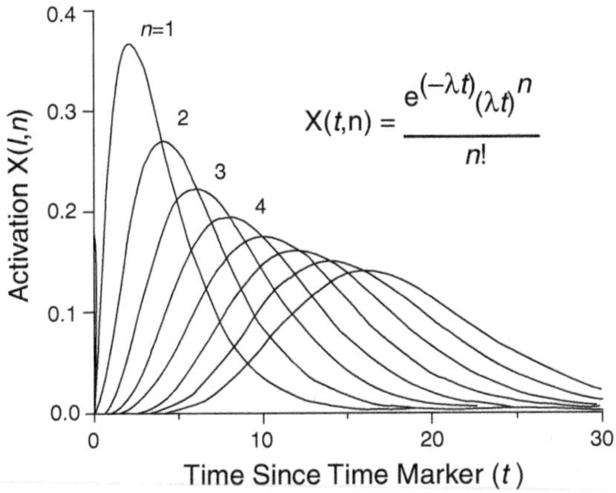

Fig. 9.5 Activation levels of LeT's states (n = 1....). Activation levels rise and fall with elapsed time, and later states have flatter activation profiles than earlier states

states are active cannot, in itself, generate the scalar properties of time, so to accomplish this, the rate of transition from one state to another is linked to the overall rate of reinforcement, with faster transitions occurring with higher rates of reinforcement. This solution is similar to that used by BeT to derive scalar timing from Poisson-like processes. A later version of LeT (Machado et al., 2009) changed the temporal distribution of the states so that the scalar property would be easier to derive. The "states" in LeT are less closely identified with potentially observable adjunctive behaviours than those in BET, and may simply be theoretical constructs, like the "ticks" of SET's internal clock. In addition to periods of maximal activation, each state has associative links to the observed operant behaviour—for example, lever-pressing or key-pecking—and these associative links can be modified. If a reinforced response occurs during state X, then the links between the operant response and state X are strengthened. If a response occurs during state X, and the response is not reinforced, the associative link between the state and the measured response is weakened. If a state is not active when a reinforcer is delivered, its links with the operant response remain unchanged.

Suppose that the animals who will later be exposed to FI initially respond regardless of time elapsed since reinforcement, so all states occurring during the interval are, to some extent, associated with the measured response. We can easily see how the normal pattern of FI responding can be acquired. States active at the time of reinforcement will have their links to the operant response strengthened, whereas those active early in the interval will have their association with the response weakened, as reinforcers are never delivered early in the interval, even if responses occur. Thus the pattern of behaviour under FI that is typically observed, of little or no responding early in the interval, with increasing average response rates later (see Fig. 9.1), will be acquired. LeT, unlike SET or BeT, incorporates a model of learning in its structure, and can therefore account for acquisition of and change in response patterns, in contrast to the other theories (see Machado, 1997, for some examples).

LeT's learning rules make some predictions beyond the scope of SET and BeT. For example, consider a mixed FI schedule where reinforcers are available for responses either according to FI A or FI B (where B > A), with nothing signalling which particular requirement is in force on any given interval. LeT predicts the overall pattern of responding on the longer (FI B) intervals. Response rates will increase from low levels early in the interval to an initial peak at A, then decline, only to rise to another peak at B. This is because states active at both A and B become associated with reinforcement, as responses are reinforced at both times. If A and B are sufficiently different that they have few overlapping states, responding will decline between A and B, as states which are active at these times lose their association with the operant response, because it is never reinforced at these times. LeT can go further. Suppose that the probability of reinforcement at A is changed. LeT predicts that the more often responses are reinforced at time A, the higher the response rate at A will be, and this is true, as shown by the data in Fig. 9.6, taken from Whitaker, Lowe, and Wearden (2008).

However, the ability to predict response rates on mixed FI schedules also creates a problem for LeT. Suppose that responses are reinforced equally often at times A and B. LeT predicts that the response rate at the time of reinforcement of the longer interval (B) should always be higher than the rate at the time of the shorter interval (A). Why? On an interval

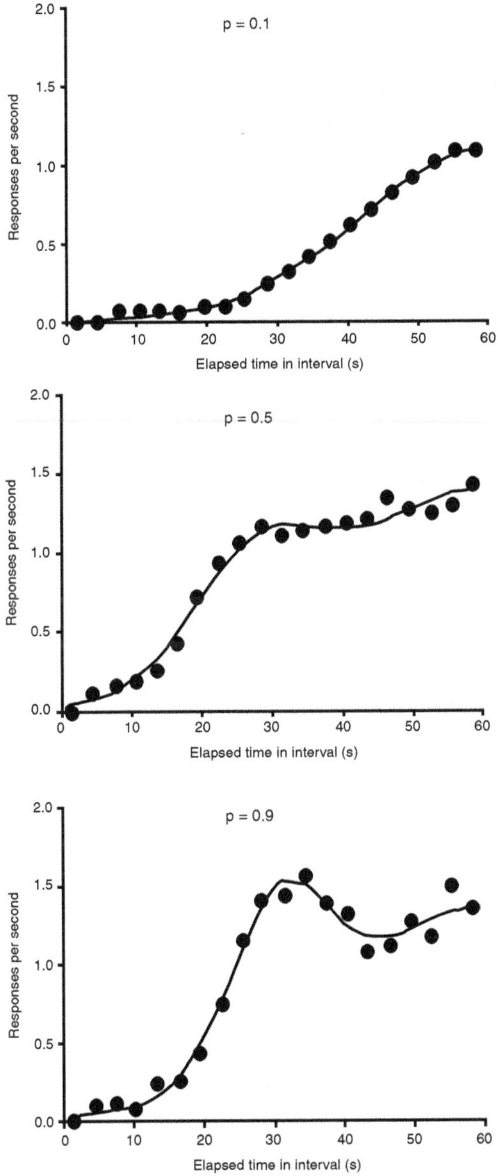

Fig. 9.6 Data from Whitaker et al. (2008). The data are the response rates versus elapsed time in FI 60-s components of a mixed FI 30-s/FI 60-s schedule. In the different panels, the probability of reinforcement in the FI 30-s component was varied over values of 0.1, 0.5, and 0.9

where responses are reinforced according to FI A, then, the states usually active at B have not yet occurred, so their association with the response does not change. On an FI B interval, however, the operant response will be not be reinforced at time A, so the association with the operant response for states active at A will be weakened by non-reinforcement. In contrast, the states active at B will always be associated with reinforcement, and thus will have maximal strength and, consequently, higher response rates.

Whitaker et al. (2003), as mentioned earlier, conducted an extensive study of the performance of rats on two-value mixed FI schedules. The authors found many cases where the peak rate of response at the shorter FI value of the mixed schedule was higher than at the longer value, although cases where the rate was higher at the longer value did sometimes occur, consistent with LeT's predictions (see Fig. 9.2, for example). The issue of response rate on FI and related schedules is given further treatment in Wearden and Lejeune (2006).

LeT's treatment of bisection illustrates some important differences with SET which may not be immediately obvious from my brief description of its structure. Suppose we have a standard bisection task for pigeons. The centre key of a three-key operant chamber is illuminated with white light for either 1 or 4 s, and then side keys are lit, one red and one green. A choice of red is rewarded if the centre key was illuminated for 1 s; a choice of green if the centre key was illuminated for 4 s. LeT assumes that associations are formed between the states active at the end of the centre key stimulus and the different responses (peck red, peck green). Obviously, earlier states will tend to be associated with red (1-s signal) and later states with green (4-s), so the normal result of an increasing number of LONG/green choice responses as the duration of illumination of the centre key increases is to be expected. However, there is another, less immediately clear feature of the change in the link between states and the two choices, and this is what Machado et al. (2009) call "context dependency." In the example above, when a response to red is reinforced, links between the states active at this time will *increase* their association with pecking red, but *decrease* their association with pecking green. Conversely, states active when a response to green is rewarded will have that association strengthened, but their association with pecking red weakened. This means that the changes in the association of states with the different operant responses (peck red and peck green) are dependent on the reward contingencies for *both* responses. I have laboured this point because it becomes important for understanding differences between SET's and LeT's predictions on a task frequently used by Machado and colleagues, and which is claimed to distinguish critically between SET and LeT, that of *double bisection.*

Double bisection (Arantes & Machado, 2008; Machado & Keen, 1999) proceeds in three phases. Pigeons are taught a bisection task with 1-s and 4-s samples associated with responses to red and green keys, as in the example above. They are also taught a second bisection discrimination, with 4-s and 16-s samples, but this time the reinforced responses are pecking a blue key after 4 s and pecking a yellow key after 16 s. When these two discriminations have been mastered, the sample key duration is varied over values from 1 to 16 s, and this time the choice is between green and blue keys. Which does the animal peck as the sample duration increases, and why? In fact, the animal shows an increasing preference for pecking the green key as the sample duration increases. This might seem strange, for two reasons. One is that the choice between green and blue keys is novel; the other is that responses to the green and blue keys have both been rewarded equally often after the same sample duration, 4 s.

The explanation of this effect according to LeT is rather complex, and here I follow the version given by Machado et al. (2009). Suppose we have a succession of states, which we can classify as *early* (e.g. around 1 s), *middle* (around 4 s), and *late* (around 16 s). After training on the 1-s and 4-s bisection task, the peck red response will be strongly associated with early states, will have zero or very weak associations with middle states, and will have an intermediate association with late states. This latter condition arises because late states have not yet occurred, and so they cannot change their associations with red regardless of whether responses to red are reinforced. For the peck green response, the early states have near zero association, the middle states a strong association, and the late states some intermediate association, as they do for the peck red response. An additional factor for peck red and peck green is that neither response has been contrasted with a response reinforced when late states were active, the peck yellow response. Recall that the "context dependency" feature of

LeT is contingent on the reinforcement of one or the other of two competing responses, whereas pecks to red and green are never competitors to pecks to blue and yellow during training.

If we now look at the peck blue and peck yellow responses, these two responses have intermediate associations with early states, and blue has a strong association with middle states, but near zero association with late states. For yellow, the reverse is true: peck yellow has near zero association with middle states (as pecks to blue were reinforced after a 4-s sample), but strong associations with late states. The outcome of this process is that blue is preferred at short samples (less than 4 s), preference is about equal at around 4 s, and preference for green then increases as the sample stimulus increases in length, which is the result obtained.

Machado et al. (2009) argue that this change in preference cannot be accounted for by SET, because essentially, the two responses (peck green and peck blue) have been associated equally often with the same 4-s sample, and thus should produce equal numbers of responses regardless of sample duration. The main difference between SET's and LeT's predictions is that SET predicts a flat preference function for peck green as sample duration changes, whereas LeT predicts the pattern of increasing green preference found in data, although there are other differences in predicted behaviour which are less marked but generally supportive of LeT over SET.

The reader interested in all the details of this issue should consult Machado et al. (2009), but in essence, the difference in predictions between SET and LeT is related to the "context-dependent" aspects of the way that associations between states and responses are supposed to change according to LeT, and the presence of some pre-existing links between states and responses. SET does not have this context-dependent aspect, and so is unable to deal with the results of the double bisection task without some modification.

Machado et al. (2009) also address some problems that LeT encounters with mixed FI schedules, such as its prediction that the timing of the two components will not be scalar, and that response rates are always higher at the time of the longer versus the shorter interval, when the longer interval is in force. Both these predictions are false: timing on mixed FI is usually scalar, and response rates are often higher at the time of the shorter interval (Whitaker et al., 2008). The details of the modifications to LeT are beyond the scope of the present discussion, but Machado et al. (2009) provide them. The changes improve LeT data fitting, but some problems, particularly those related to mixed FI schedules, remain.

In conclusion, Machado's LeT is a bold attempt to marry a model of response learning with a model of timing. As such, it is a unique achievement. SET cannot predict response rates without some additional features (Wearden & Lejeune, 2006, provide some ideas as to how this could be achieved), and cannot deal with data from the double bisection procedure without modification. LeT does not produce a completely correct account of all the phenomena it treats, but in this respect it does not differ from BeT or SET. Like BeT, a potential weakness is that LeT may be difficult to seriously consider as an account of human timing, although its states are largely hypothetical, and thus might occur in situations where humans time brief intervals, a situation virtually impossible for BET to deal with. However, a study of temporal bisection in humans (Jozefowiez, Polack, Machado, & Miller, 2014), reported that data from an experiment in which the probability of the standard *Short* and *Long* durations was manipulated were consistent with ideas from the Behavioural Economic Model (Jozefowiez, Staddon, & Cerutti, 2009), which shares some principles with LeT, in particular the use of similar rules for associating responses with outcomes. It was also acknowledged that some SET-consistent accounts of bisection in humans (like that of Wearden, 1991b; see Chap. 4) might be compatible with the effects obtained.

Summary

The main aim of this chapter has been to discuss some of the principal behavioural results obtained from animal timing experiments, and in particular, to show how three theoretical approaches—SET, BeT, and LeT—have explained many of them. The consistency of the data and reproducibility of results from animal timing experiments has encouraged other theories (see, for example, Staddon & Higa, 1999, and Dragoi, Staddon, Palmer, & Cerutti, 2003). However, the three theories reviewed here are by far the most highly developed and most frequently applied to data. BeT and LeT are also interesting because of their metatheoretical position linked to Skinnerian behaviourism, which tries to reduce explanatory reliance on internal processes of the sort that SET freely uses. Although the cognitive "clockwork" of SET may seem completely natural as an explanatory system—particularly to those largely concerned with psychological processes in humans, which are almost always proposed to be explained by some sort of cognitive process—LeT and, particularly, BeT are reminders that ingenious theorists can often explain experimental results in other ways. As mentioned above, BeT and LeT may encounter serious problems when applied to the timing of short durations by humans, as well as "one-shot" discrimination of time intervals never presented before, but they are both more than worthy challengers to SET in the domain of animal timing.

Appendix: Correlations Between Performance Measures from the Peak Procedure

The aim of the arguments advanced in Gibbon and Church (1990) and Church et al. (1994) was to determine the contribution of different sources of variance to performance on individual peak procedure trials. In SET, there can be trial-to-trial variance in both memory/timing and the threshold for starting and stopping responding, as discussed in Chap. 3 and earlier in this chapter. A simplified outline of how these might contribute to performance is given here, but for full mathematical exposition, the reader should consult the articles mentioned immediately above.

In any attempt to assess the contribution of different variance sources, the data of interest are the correlations between the different performance measures (mainly start, stop, and spread). Gibbon and Church (1990) and Church et al. (1994; see their Table 2, p. 146) characterize the correlation pattern as follows, in which the values given are the means from Church et al.'s own study. Start and stop values are positively correlated (mean value 0.31), whereas start and spread (the length of the response period) are negatively correlated (mean -0.33).

Equation (9.1), the usual SET rule for judging the similarity of two time values, determines the start time and the stop time. Suppose that *S* is the standard FI value of the peak procedure task. The start of responding (*ST*) occurs when

$$(S - ST) / S < b$$

Or, rearranging, $ST = S - Sb$

And the stop (*SP*) occurs when

$$(SP - S) / S > b, \text{or} SP = S + Sb$$

where *b* is the threshold.

We consider first the case where there is only trial-by-trial variability in *S*, and *b* remains constant. Suppose that the FI value in the peak procedure is 60 s, and the threshold is 0.25. Thus, $ST = 45$ s, $SP = 75$ s, and the spread is $75 - 45 = 30$ s. The case just discussed assumes that the FI value is represented correctly, but SET proposes the existence of memory/timing variance, so on another trial, the value sampled from the memory might be "short," 50 s. Now, $SP = 48.5$ s, $ST = 62.5$ s, and the spread is 24 s. On another trial, a "long" 70-s value might be sampled, so $ST = 53.5$ s, $SP = 87.5$ s, and the spread is 34 s. From these three examples, we can see that "early" starts lead to "early" stops, and "late" starts lead to "late" stops, so *ST* and *SP* are positively correlated. However, *ST* and the spread are also positively correlated: a "late" start gives rise to a longer response period than an "early" start. In data, ST and *SP* are positively correlated, but *ST* and the spread are *negatively* correlated. This shows that an account with memory/timing variance alone cannot produce the full pattern of correlations obtained.

Now, consider the effect of trial-by-trial variability only in the threshold, *b*, with *S* constant at 60 s on each trial. If $b = 0.25$, we have the case above with $ST = 45$ s, $SP = 75$ s, and spread = 30 s. If the threshold is more conservative ($b = 0.2$), then *ST* is 48 s, *SP* is 72 s, and the spread is 24 s. This is intuitively reasonable: if the decision to start and stop responding is more conservative, the response period should start later and finish earlier. In contrast, if $b=0.3$, then $ST=42$ s, $SP=78$ s, and the spread is 36 s. Thus, when the threshold alone is varied, early starts lead to late stops, and late starts to early stops, so *ST* and *SP* are *negatively* correlated. However, the examples also show that early starts lead to longer response periods (spreads) than later starts: in other words, start and spread are *negatively* correlated. By itself, this model does not fit overall correlation patterns, as *ST* and *SP* are positively correlated in data.

The obvious suggestion that can be derived from the outline above is that an account with both memory/timing variance and threshold variance is needed to model the pattern of correlations in data, and this is exactly what Gibbon and Church (1990) and Church et al. (1994) propose. An additional issue is the number of samples of both memory and threshold used in a single trial. For example, the animal might sample the reference memory once and the threshold once, so the same threshold is used for start and stop decisions. Alternatively, the reference memory might be sampled twice (once for the start decision and once for the stop decision), and likewise, the threshold might be sampled twice for the same decisions. The arguments are too complex to illustrate here, but the articles cited above suggest that a single memory sample per trial is used, with either one threshold or different thresholds for start and stop.

10

Methods Commonly Used in Time Perception Research

Temporal Reproduction

In temporal reproduction, participants produce a motor response which is intended to reflect the duration of a previously presented stimulus or event. There are various ways of doing this. In early research by Vierordt (1868), for example, two taps were made on a glass plate, with some interval between them, and the participant made a single tap such that the time between the second tap on the plate and the participant's response was the same as the time between the two taps. This may sound like a rather crude method, but in fact it is in principle very similar to the "ready-set-go" procedure used in more recent research (Jazayeri & Shadlen, 2010). The interval to be reproduced is the time between a "Ready" and "Set" cue, and the participant must then respond once such that the time between the "Set" and the response is the same as that between the two previously presented cues. In Jazayeri and Shadlen's experiment, feedback was provided as to reproduction accuracy. If the interval reproduced was within some time window around the target, a green signal was then presented, although in many reproduction experiments, no feedback is given with regard to performance accuracy.

J. Wearden, *The Psychology of Time Perception*,
DOI 10.1057/978-1-137-40883-9_10

Another method involves the presentation of a stimulus lasting for some target duration, followed by a short delay, at which point the participant initiates and terminates an interval (by pressing twice on a response key), or holds down a response key, so that the time between the responses or the duration for which the key is depressed is the same as that of the target stimulus. A variant of this method involves presenting two stimuli, in which the first defines the duration to be reproduced. Following termination of the first stimulus, a gap ensues, the stimulus is then presented again, and it is terminated by a response from the participant when it is judged to have reached the same duration as the first stimulus.

In temporal reproduction, durations are usually presented in random order. If the same duration is presented repeatedly, and participants know this, it is possible that they will simply try to reproduce their previous responses, turning the task into a form of interval production.

Temporal reproduction is obviously limited at the lower end by reaction time considerations: a person cannot reasonably be expected to reproduce an interval much shorter than 300 ms. However, it can be used without difficulty for multi-second durations (e.g. Boltz, 1994), provided that the participant can maintain attention to the task if the duration to be reproduced is of considerable length. Reproduction can also be used in both prospective and retrospective timing studies. A person can be instructed simply to attend to a stimulus or other event, then unexpectedly be asked to reproduce its duration, thus providing a retrospective time judgement (as in Boltz, 1994).

The normal measures of performance on reproduction tasks are the mean interval reproduced, and some measure of variability, such as the standard deviation or the coefficient of variation. If a range of intervals is reproduced, it is common to plot measures such as the mean interval reproduced against the target interval. A frequent result in reproduction, albeit one that is not always found, is conformity to Vierordt's Law (see Lejeune & Wearden, 2009, for discussion). This is the phenomenon by which short time intervals tend to be reproduced as longer than they really are, whereas longer intervals tend to be reproduced as shorter than their real value, with an "indifference point," where the interval is correctly reproduced on average, being somewhere in the range. Lejeune and Wearden (2009) discuss this issue more fully.

A variant of reproduction which has attracted interest in recent years is the "human peak-interval" procedure developed by Rakitin et al. (1998). This method was intended to provide an analogue for humans of the peak procedure developed in animal timing studies by Roberts (1981) from earlier work by Catania (1970). The use of the peak procedure in animal timing is discussed in Chap. 9. In a version for humans, the participant receives initial presentations of a visual stimulus, which begins as one colour (e.g. blue), and after some target length of time changes colour (e.g. to magenta). Participants are instructed to observe the colour change and to try to judge when it occurs. On the critical "peak trials," which follow this training, the stimulus will remain blue, and the participant's task is to produce a response which "brackets" the target time. For example, the instructions might be to begin making repeated responses just before the colour change is expected, and to stop afterwards. In a variant (Malapani, Deweer, & Gibbon, 2002), individuals were instructed to depress a spacebar just before the expected moment of colour change and to release it when the moment of colour change was judged to have passed. The measure of responding in this case is the relative frequency or rate of response during the peak trials as a function of time elapsed in the trial. Obviously, the specific instructions given to participants in the peak trial are critical. If the target time is 6 s, for example, then responding from the beginning of the trial or holding down the spacebar from the beginning of the trial (strategies that would guarantee responding at the target time) must be discouraged, and this is often accomplished by some kind of response feedback (see Rakitin et al., 1998, for details).

Although it seems obvious that motor factors may play an important role in temporal reproduction, particularly when short durations are reproduced, motor factors have received little experimental and theoretical attention in discussions of reproduction. The exception is the study by Droit-Volet (2010) using both children and adults, in which the author found that the reproduction of what in her study was a short duration (2.5 s) was highly correlated with an individual's reaction time. For example, children 5 and 8 years of age reproduced the 2.5-s duration as longer than it really was, and longer than the adults' reproductions, but their reaction times were also longer, so it is possible that the children's reproductions were affected by the longer time they required to produce a response.

Few models of reproduction exist, and none seem satisfactory. Wearden (2003) simulated Vierordt's Law in reproduction, assuming an interaction of an internal timing process and response time, but this model could not account for Vierordt's Law when intervals of more than a few seconds were reproduced (see Lejeune & Wearden, 2009, for examples showing that this can happen). Wackerman and Ehm (2006) simulated reproduction using a complex "dual clepsydra" model, which predicted that reproductions will always be shorter than the target time, although they progressively undershoot it as the target time lengthens. Once again, this is in accord with some data, although by no means all.

Interval Production

In interval production, participants are verbally instructed to produce a certain target interval, e.g. 5 s or 500 ms. They do this by emitting some response with the required temporal characteristics, for example, pressing a button twice to start and stop the interval (Wearden & McShane, 1988) or holding down a spacebar for the required length of time (Wearden, Wearden, & Rabbitt, 1997). In a variant of this procedure, the interval is started by a signal (such as a click or flash), and the participant waits for the required time before responding once (Penton-Voak, Edwards, Percival, & Wearden, 1996), resulting in a "production-by-waiting" method. The advantage with this approach is that the moment of the start of interval production is controlled. Like reproduction, interval production cannot be used with very short durations, but can be used with long durations (e.g. Craik & Hay, 1999). Feedback as to the interval produced may or may not be given, and trials with the same interval to be produced are sometimes given in blocks (as in Wearden & McShane, 1988). Unlike reproduction, interval production cannot be used in retrospective timing studies, as participants are informed in advance as to the duration they must produce, making the timing prospective by definition.

As in the case of reproduction, the typical measures of behaviour recorded are the mean interval produced and some variability measure, often plotted against the target interval.

Feedback as to the duration actually produced can be given, although this is not always provided. Some research has investigated the effects of feedback on controlling performance (see Franssen & Vandierendonck, 2002; Ryan & Fritz, 2007), reaching the general conclusion, perhaps unsurprisingly, that both the reference memory of the target duration and decision processes governing when to respond can be affected by feedback.

Verbal Estimation of Duration

With this method, the participant estimates the duration of stimuli or other events using conventional time units such as milliseconds, seconds, or minutes. This method provides great flexibility, and can be used with a wide range of intervals, with durations ranging from milliseconds to minutes or longer, as it is not limited by the speed of motor processes, unlike reproduction or production. It can also be used for both prospective and retrospective timing. In some cases, examples of standard durations, such as 1 s, or some sort of performance-related feedback are given to participants, but this may have little effect on their performance on average (Wearden & Farrar, 2007), and is not common. The measure of performance normally recorded is simply the estimate given for each stimulus duration presented, or the mean of the estimates if a duration is presented more than once. In the latter case, measures of variability in estimates at each duration used can also be taken, such as the standard deviation. A problem with verbal estimation, potentially affecting both mean and measures of variability, is "quantization," which is the tendency for participants to use certain "rounded" values. For example, when stimulus durations less than 1 s are used, and the estimation scale is milliseconds (1000 = 1 s), then values like 250, 500, and 1000 are commonly used, whereas a value of 428 would be rare. In practice, a participant might use only ten estimate values, regardless of the range of stimuli, and sometimes fewer than that.

If the focus of interest is the estimation of only a small number of stimuli, such as three or five, and the stimuli are repeated, a potential problem is that individuals may identify the durations (particularly the shortest and longest) and always use the same estimate values for these stimuli, reducing the variability of their estimates to zero. One way of avoiding this is to embed the "target" durations (i.e. those time values which are the focus of experimental interest) amongst other random durations. For example, suppose that the values of experimental interest were 450, 750, and 1050 ms, and the experimental design required them to be repeated. These three "target" stimuli could be presented in random order in a block of five or six stimuli, where the durations of the non-target stimuli were random, that is, chosen from some duration range but otherwise unpredictable. It might also be advisable to use a random duration range exceeding the span of the targets, so that the targets are neither the longest nor shortest stimuli in the set. Thus, in the example above, target durations might be chosen from a uniform distribution ranging from 300 to 1200 ms. In this case, if we have a repeated block of five stimuli, two trials on each block are "wasted," but this may be preferable to the risk of the target stimuli being repeatedly labelled with the same estimate value.

My own personal experience also suggests that when verbal estimation is used, it is advisable to inform participants of the range of possible responses so as to avoid extreme estimates. For instance, in the example above, it is probably a good idea to inform participants that all duration values are between 300 and 1200 ms, and that only values in this range should be used. If values outside the range are used, they can then be filtered out, or data from participants using estimates outside this range can be discarded. Without some restrictive instruction of this sort, the problem arises as to how estimates outside the range (which may be highly aberrant, and thus can markedly distort the mean or variability of estimates) should be treated.

Discrimination Methods

A wide variety of discrimination methods are used in time perception research. These usually involve some kind of direct comparison between pairs of stimuli, so the judgement might involve deciding which is the longer of two presented stimulus durations, or whether the stimuli have the same duration. For example, a tone 400 ms long might be contrasted with other tones, some shorter, some longer. Ideally, the order of presentation should be randomized to avoid time–order errors (Hellstrom, 1985). A variant of this method uses trial-by-trial changes in duration values (see Rammsayer & Brandler, 2007), with the changes dependent on the accuracy of the response. Suppose one of the stimuli is 400 ms in length. On the first trial of an adaptive procedure, this might be contrasted with a stimulus which is clearly longer or shorter, such that the first response is always correct. A correct response reduces the difference between stimuli by some step value, whereas an incorrect response increases the difference between stimuli. As the gap between the two durations narrows, the step values may be reduced. Such a method has been used in many studies to determine the differential threshold for duration: the smallest difference between two durations that can be detected some percentage (such as 75 %) of the time. For example, if the upward step value (the increase in the difference when an incorrect response is made) is three times the downward step value (the decrease in the difference when a correct response is made), then the difference converges on the 75 % threshold for the stimuli used.

Bisection

Temporal bisection is usually associated with scalar timing theory (discussed in Chaps. 3 and 4), and in this context was developed from an animal experiment by Church and DeLuty (1977). However, this was predated by studies of temporal bisection in humans (e.g. Bovet, 1968).

There are a number of variants of this method, but the procedure is generally as follows. The participant initially receives examples of *Short* (*S*) and *Long* (*L*) standards—for example, tones 200 and 800 ms in duration. With human adults, a few presentations, perhaps three to five, are normally given. The *S* and *L* standards are associated with different responses (e.g. the S and L keys on a computer keyboard). In some cases, the participant's ability to discriminate the standards is tested, and a criterion (e.g. 9/10 correct responses) is used. This is common in studies with children, who may need extensive training, particularly if the *S* and *L* stimuli are similar in duration (e.g. Droit-Volet & Wearden, 2001).

When the training phase—which can be very short—is completed, comparison stimuli at intermediate points between *S* and *L* are presented, as well as *S* and *L* themselves. So, for example, the comparison stimuli in the example above might be 200, 300, 400, 500, 600, 700, and 800 ms, although the spacing need not be linear between *S* and *L* (e.g. Wearden & Ferrara, 1995). After each comparison is presented, the usual task involves deciding whether it was closer in duration to *S* (press "S") or *L* (press "L"). In some cases, however, participants are asked whether each stimulus is *S* or *L* (Allan & Gibbon, 1991), although this feasible only if *S* and *L* are very similar. Sometimes feedback is given, but it can be provided only for the stimuli that are *S* and *L*, as there is no correct answer for intermediate stimuli. In other cases (e.g. Wearden, 1991b), there is no feedback, but the standards *S* and *L* are presented again, followed by additional comparisons, and so on.

The typical measure of performance on bisection is a *psychophysical function* in the form of the proportion of Long responses (i.e. decisions that a stimulus duration is closer to *L* than *S*), plotted against stimulus duration. This usually takes the form of an increasing ogival function, progressing from few Long responses when *S* is presented to nearly 100 % when *L* is presented, at least in student-age adults. The psychophysical function can be analysed to yield a number of measures. The most important of these are (a) the *bisection point*, or the stimulus duration giving rise to 50 % "Long" responses, and (b) the Weber ratio, which is a measure of the steepness of the psychophysical function (with steeper curves producing smaller Weber ratios). The Weber ratio is calculated by taking the difference between the stimulus durations giving rise to 75 % and 25 % Long responses, halving it, then dividing by the bisection point value (see Wearden, 1991b, for examples). There are, however, other equivalent measures of the steepness of the curve, deriving from curve fitting. In general, the Weber ratio or equivalent measure is considered a measure of timing sensitivity, with low values showing high sensitivity to duration, and high values showing low sensitivity.

There are some variants of the bisection method outlined above. One of these is the *partition bisection* method, developed by Wearden and Ferrara (1995). Here, no standards are presented, in the sense that no stimuli are explicitly identified as the standards. Suppose that a duration set of 200 to 800 ms in 100-ms steps is used. The stimuli are presented in random order, and the participants' instruction is to classify them as Short or Long using any criterion they choose. The stimulus set is repeated a number of times—ten times, for example—and data from the last five blocks are taken. Results from Wearden and Ferrara (1995) suggest that this method will result in psychophysical functions which are impossible to distinguish from those produced by the "normal" bisection method, where standards are initially presented.

Another variant of bisection is a "roving standard" (or episodic) bisection method (e.g. Wearden & Bray, 2001). In Wearden and Bray's article, this involved the presentation of three stimuli on each trial. The first two were the "standards" for the trial, which are chosen randomly on each trial but which have some constant relation to one another; for example, the long stimulus is four times the length of the short stimulus. The short and long "standards" are presented in a counterbalanced order across trials. The third stimulus on the trial is the comparison, which is some proportion of the short and long standards, and the participant's task is to judge whether it is more similar to the short or long "standards" presented earlier on the trial. In this type of procedure, the "standards" change from one trial to the next, and the participant is informed of this. This means that it is unlikely that the participant uses any kind of temporal "reference memory," and relies instead on the working memory of the durations presented on each trial.

Temporal Generalization

The technique of temporal generalization for humans was developed by Wearden (1992) from an earlier animal study by Church and Gibbon (1982). Participants are initially presented with a stimulus of a "standard" duration (e.g. a tone 400 ms long) and are asked to remember the standard. Following this training, which can be very brief for student-age participants, comparisons are presented (e.g. tones lasting 100, 200, 300, 400, 500, 600, and 700 ms, in the example above), and the participant must decide whether each tone was or was not the standard duration, using a YES or NO response. In temporal generalization, unlike bisection, there is always a correct response to each stimulus presented, so feedback as to correctness can be given after each response. In some cases, however, no feedback is given, but the standard is periodically presented again.

The typical way to represent temporal generalization performance is to construct a *temporal generalization gradient*, which is the proportion of YES responses (decisions that a stimulus duration was the same as the standard) plotted against comparison stimulus duration. In student-age adults, such gradients usually (a) peak at the standard and (b) are skewed to the right, so stimuli longer than the standard by a certain amount give rise to more YES responses than stimuli shorter than the standard by the same amount. For example, Wearden (1992) found that participants confused a 500-ms stimulus more often than a 300-ms stimulus with a 400-ms standard, although this is not always observed. The width of the temporal generalization gradient can also be used to provide a measure of performance variability (e.g. Wearden et al., 1997), although many studies of temporal generalization in humans use theoretical models (discussed in Chap. 4) to derive underlying measures of performance.

Comparison stimuli may be linearly or logarithmically spaced around the standard, or may have biased distribution, for example, with a greater number of stimuli longer than the standard than shorter, or vice versa.

An "episodic" variant of temporal generalization which is not dependent on retention of a standard duration was developed by Wearden and Bray (2001). On each trial, two stimuli are presented. One (the "standard") is a random value drawn from some distribution (e.g. 300–500 ms), and the other (the "comparison") is some multiple of this "standard." For example, multiples might range from 0.25 to 1.75 of the "standard." The "standard" and "comparison" durations vary from trial to trial, and the order of "standard" and "comparison" stimuli is counterbalanced over trials. The participant's task is to decide whether the two stimuli on the trial have the same duration (responding YES or NO, as in the typical temporal generalization method). The proportion of YES responses is plotted against the "comparison/standard" multiple (in the example above, from 0.25 to 1.75) to produce a temporal generalization gradient. Stimuli from different duration ranges can be intermixed with this method; that is, standards drawn from a 300- to 500-ms distribution and their associated comparisons can be intermixed with those drawn from another distribution, and the different temporal generalization gradients plotted separately, as in Wearden and Bray (2001, Fig. 4).

References

Alderson, M. J. (1974). Effect of increased body temperature on the perception of time. *Nursing Research, 23*, 42–49.

Allan, L. G. (1979). The perception of time. *Perception and Psychophysics, 26*, 340–354.

Allan, L. G. (2002). Are the referents remembered in temporal bisection? *Learning and Motivation, 33*, 10–31.

Allan, L. G., & Gibbon, J. (1991). Human bisection at the geometric mean. *Learning and Motivation, 22*, 39–58.

Anderson, J. R., Bothell, D., Byrne, M., Douglass, D., Lebiere, C., & Qin, Y. (2004). An integrated theory of mind. *Psychological Review, 111*, 1036–1060.

Angrilli, A., Cherubini, P., Pavese, A., & Manfredini, S. (1997). The influence of affective factors on time perception. *Perception and Psychophysics, 59*, 972–982.

Arantes, J., & Machado, A. (2008). Context effects in a temporal discrimination task: Further tests of the scalar expectancy theory and learning-to-time models. *Journal of the Experimental Analysis of Behavior, 90*, 33–51.

Arlin, M. (1986). The effects of quantity and depth of processing on children's time perception. *Journal of Experimental Child Psychology, 42*, 84–98.

Arstila, V. (2012). Time slows down during accidents. *Frontiers in Psychology, 3*, 1–9.

J. Wearden, *The Psychology of Time Perception*,
DOI 10.1057/978-1-137-40883-9

Baddeley, A. (1966). Time-estimation at reduced body-temperature. *American Journal of Psychology, 79*, 475–479.

Bausenhart, K. M., Dyjas, O., & Ulrich, R. (2014). Temporal reproductions are influenced by an internal reference: Explaining the Vierordt effect. *Acta Psychologica, 147*, 60–67.

Beck, A., Steer, R. A., & Carbin, M. G. (1988). Psychometric properties of the Beck depression inventory: Twenty-five years of evaluation. *Clinical Psychology Review, 8*, 77–100.

Bell, C. R. (1965). Time estimation and increases in body temperature. *Journal of Experimental Psychology, 70*, 232–234.

Bell, C. R. (1966). Control of time estimation by a chemical clock. *Nature, 210*, 1189–1190.

Bell, C. R. (1975). Effects of lowered temperature on time estimation. *Quarterly Journal of Experimental Psychology, 27*, 531–538.

Bell, C. R., & Provins, K. A. (1963). Relation between physiological responses to environmental heat and time judgments. *Journal of Experimental Psychology, 66*, 572–579.

Birngruber, T., Schroter, H., & Ulrich, R. (2014). Duration perception of auditory and visual oddball stimuli: Does judgment task modulate the temporal oddball effect? *Attention, Perception, and Psychophysics, 76*, 814–828.

Block, R. A. (1974). Memory and the experience of duration in retrospect. *Memory and Cognition, 2*, 153–160.

Block, R. A. (1978). Remembered duration: Effect of event and sequence complexity. *Memory and Cognition, 6*, 320–326.

Block, R. A., Hancock, P. A., & Zakay, D. (2010). How cognitive load affects duration judgments: A meta-analytic review. *Acta Psychologica, 134*, 330–343.

Block, R. A., & Reid, M. A. (1978). Remembered duration: Evidence for a contextual change hypothesis. *Journal of Experimental Psychology: Human Learning and Memory, 4*, 656–665.

Block, R. A. (1992). Prospective and retrospective duration judgment: The role of information-processing and memory. In F. Macar, V. Pouthas, & W. J. Friedman (Eds.), *Time, action, and cognition* (pp. 141–152). Dordrecht: Kluwer.

Block, R. A., & Zakay, D. (1997). Prospective and retrospective duration judgments: A meta-analytic review. *Psychonomic Bulletin and Review, 4*, 184–197.

Block, R. A., Zakay, D., & Hancock, P. A. (1998). Human aging and duration judgments: A meta-analytic review. *Psychology and Aging, 13*, 584–596.

Boltz, M. G. (1993). Time estimation and expectancies. *Memory and Cognition, 21*, 853–863.

Boltz, M. G. (1994). Changes in internal tempo and effects on the learning and remembering of event durations. *Journal of Experimental Psychology: Learning, Memory, and Cognition, 20*, 1154–1171.

Boltz, M. G. (1998). Task predictability and remembered duration. *Perception and Psychophysics, 60*, 768–784.

Boltz, M. G. (2005). Duration judgments of naturalistic events in the auditory and visual modalities. *Perception and Psychophysics, 67*, 1362–1375.

Bovet, P. (1968). Echelles subjectives de durée obtenues par une méthode de bissection. *L'Année Psychologique, 68*, 23–36.

Bradley, M. M., & Lang, P. J. (1999). *International affective digitized sounds (IADS): Stimuli, instruction manual and affective ratings* (Tech. Rep. No. B-2). Gainesville, FL: The Center for Research in Psychophysiology, University of Florida.

Brown, S. W. (1985). Time perception and attention: The effects of prospective versus retrospective paradigms and task demands on perceived duration. *Perception and Psychophysics, 38*, 115–124.

Brown, S. W. (1997). Attentional resources in timing: Interference effects in concurrent temporal and nontemporal working memory tasks. *Perception and Psychophysics, 59*, 1118–1140.

Brown, S. W., Collier, S. A., & Night, J. C. (2013). Timing and executive resources: Interference patterns between temporal production and shifting, updating, and inhibition tasks. *Journal of Experimental Psychology: Human Perception and Performance, 39*, 947–963.

Brown, G. D. A., McCormack, T., Smith, M., & Stewart, N. (2005). Identification and bisection of temporal durations and tone frequencies: Common models for temporal and non-temporal stimuli. *Journal of Experimental Psychology: Human Perception and Performance, 31*, 919–938.

Burle, B., & Casini, L. (2001). Dissociation between activation and attention effects in time estimation: Implications for clock models. *Journal of Experimental Psychology: Human Perception and Performance, 27*, 195–205.

Catania, A. C. (1970). Reinforcement schedules and psychophysical judgments. In W. N. Schoenfeld (Ed.), *The theory of reinforcement schedules* (pp. 1–42). New York: Appleton-Century-Crofts.

Church, R. M., & Deluty, M. Z. (1977). Bisection of temporal intervals. *Journal of Experimental Psychology: Animal Behavior Processes, 3*, 216–228.

Church, R. M., & Gibbon, J. (1982). Temporal generalization. *Journal of Experimental Psychology: Animal Behavior Processes, 8*, 165–186.

Church, R. M., Meck, W. H., & Gibbon, J. (1994). The application of scalar timing theory to individual trials. *Journal of Experimental Psychology: Animal Behavior Processes, 20*, 135–155.

Cohen, M. J. (1997). *Examiner's manual: Children's memory scale*. San Antonio, TX: Harcourt Brace.

Corish, D. (2005). McTaggert's argument. *Philosophy, 80*, 77–99.

Craik, F. I. M., & Hay, J. F. (1999). Aging and judgments of duration: Effects of task complexity and method of estimation. *Perception and Psychophysics, 61*, 549–560.

Craik, F. I. M., & Lockhart, R. S. (1972). Levels of processing: A framework for memory research. *Journal of Verbal Learning and Verbal Behavior, 11*, 671–684.

Creelman, C. D. (1962). Human discrimination of auditory duration. *Journal of the Acoustical Society of America, 34*, 582–593.

Crépault, J. (1979). Influence du repérage sur la durée. Etude génétique des inférences cinématiques. *L'Année Psychologique, 79*, 43–64.

Dainton, B. (2014). The phenomenal continuum. In V. Arstila & D. Lloyd (Eds.), *Subjective time: The philosophy, psychology, and neuroscience of temporality* (pp. 101–137). Cambridge, MA: The MIT Press.

Delgado, M. L., & Droit-Volet, S. (2007). Testing the representation of time in reference memory in the bisection and generalization task: The utility of a developmental approach. *Quarterly Journal of Experimental Psychology, 60*, 820–836.

Denner, B., Wapner, S., & Werner, H. (1964). Rhythmic activity and the discrimination of stimuli in time. *Perceptual and Motor Skills, 19*, 723–729.

Domjan, M. (1993). *The principles of learning and behavior*. Pacific Grove, CA: Brooks/Cole.

Dragoi, V., Staddon, J. E. R., Palmer, R. G., & Buhusi, C. V. (2003). Interval timing as an emergent learning property. *Psychological Review, 110*, 126–144.

Droit-Volet, S. (2002). Scalar timing in temporal generalization in children with short and long stimulus durations. *Quarterly Journal of Experimental Psychology, 55A*, 1193–1209.

Droit-Volet, S. (2003). Alerting attention and time perception in children. *Journal of Experimental Child Psychology, 85*, 372–384.

Droit-Volet, S. (2008). A further investigation of the filled duration illusion with a comparison between children and adults. *Journal of Experimental Psychology: Animal Behavior Processes, 34*, 400–414.

Droit-Volet, S. (2010). Stop using time reproduction tasks in a comparative perspective without further analyses of the role of the motor response on the temporal performance: The case of children. *European Journal of Cognitive Psychology, 22*, 130–148.

Droit-Volet, S., Bigand, E., Ramos, D., & Bueno, J. L. O. (2010). Time flies with music whatever its emotional valence. *Acta Psychologica, 135*, 226–232.

Droit-Volet, S., Brunot, S., & Niedenthal, P. M. (2004). Perception of the duration of emotional events. *Cognition and Emotion, 18*, 849–858.

Droit-Volet, S., Clément, A., & Fayol, M. (2003). Time and number discrimination in a bisection task with a sequence of stimuli: A developmental approach. *Journal of Experimental Child Psychology, 84*, 63–76.

Droit-Volet, S., Clément, A., & Fayol, M. (2008). Time, number and length: Similarities and differences in discrimination in children and adults. *Quarterly Journal of Experimental Psychology, 61*, 1827–1846.

Droit-Volet, S., Clément, A., & Wearden, J. H. (2001). Temporal generalization in 3- to 8-year-old children. *Journal of Experimental Child Psychology, 80*, 271–288.

Droit-Volet, S., Fayolle, S. L., & Gil, S. (2011). Emotion and time perception: Effects of film-induced mood. *Frontiers in Integrative Neuroscience, 5*, 33.

Droit-Volet, S., & Meck, W. H. (2007). How emotions colour our perception of time. *Trends in Cognitive Sciences, 1*, 504–513.

Droit-Volet, S., Mermillod, M., Concenas-Silva, R., & Gil, S. (2010). The effect of expectancy of a threatening event on time perception in human adults. *Emotion, 10*, 908–914.

Droit-Volet, S., Ramos, D., Bueno, J. L. O., & Bigand, E. (2013). Music, emotion and time perception: The influence of subjective emotional valence and arousal? *Frontiers in Psychology, 4*, 417.

Droit-Volet, S., & Rattat, A.-C. (1999). Are time and action dissociated in young children's time estimation? *Cognitive Development, 14*, 573–595.

Droit-Volet, S., Tourret, S., & Wearden, J. (2004). Perception of the duration of auditory and visual stimuli in children and adults. *Quarterly Journal of Experimental Psychology, 57*, 797–818.

Droit-Volet, S., & Wearden, J. H. (2001). Temporal bisection in children. *Journal of Experimental Child Psychology, 80*, 142–159.

Droit-Volet, S., & Wearden, J. (2002). Speeding up an internal clock in children? Effects of visual flicker on subjective duration. *Quarterly Journal of Experimental Psychology, 55B*, 193–211.

Droit-Volet, S., & Wearden, J. H. (2015). Experience sampling methodology reveals similarities in the passage of time between young and elderly adults. *Acta Psychologica, 156*, 77–82.

Droit-Volet, S., Wearden, J., & Delgado-Yonger, M. (2007). Short-term memory for time in children and adults: A behavioral study and a model. *Journal of Experimental Child Psychology, 97*, 246–264.

Droit-Volet, S., Wearden, J. H., & Zélanti, P. (2015). Cognitive abilities required in time judgement depending on the temporal task used: A comparison of children and adults. *Quarterly Journal of Experimental Psychology, 68*, 2216–2242.

Droit-Volet, S., & Zélanti, P. (2013a). Development of time sensitivity: Duration ratios in time bisection. *Quarterly Journal of Experimental Psychology, 66*, 271–286.

Droit-Volet, S., & Zélanti, P. (2013b). Development of time sensitivity and information-processing speed. *PLoS ONE, 8*(8), e71424.

Effron, D., Niedenthal, P., Gil, S., & Droit-Volet, S. (2006). Embodied temporal perception of emotion. *Emotion, 6*, 1–9.

Fechner, T. G. (1860). *Elemente der Psychophysik*. Leipzig: Breitkopf & Haertel.

Ferrara, A., Lejeune, H., & Wearden, J. H. (1997). Changing sensitivity to duration in human scalar timing: An experiment, a review, and some possible explanations. *Quarterly Journal of Experimental Psychology, 50B*, 217–237.

Ferster, C. B., & Skinner, B. F. (1957). *Schedules of reinforcement*. New York: Appleton-Century-Crofts.

Fetterman, J. G., & Killeen, P. R. (1995). Categorical scaling of time: Implications for clock-counter models. *Journal of Experimental Psychology: Animal Behavior Processes, 21*, 43–63.

Filippopoulos, P. C., Hallworth, P., Lee, S., & Wearden, J. H. (2013). Interference between auditory and visual duration judgements suggests a common code for time. *Psychological Research, 77*, 708–715.

Flaherty, M. G. (1991). The perception of time and situated engrossment. *Social Psychology Quarterly, 54*, 76–85.

Flaherty, M. G. (1993). Conceptualizing variation in the experience of time. *Sociological Enquiry, 63*, 394–405.

Flaherty, M. G. (2003). Time work: Customizing temporal experience. *Social Psychology Quarterly, 66*, 17–33.

Flaherty, M. G., Freidin, B., & Sautu, R. (2005). Variation in the perceived passage of time: A cross-national study. *Social Psychology Quarterly, 68*, 400–410.

Flaherty, M. G., & Meer, M. D. (1994). How time flies: Age, memory, and temporal compression. *Sociological Quarterly, 35*, 705–721.

Fortin, C., & Breton, R. (1995). Temporal interval production and processing in working memory. *Perception and Psychophysics, 57*, 203–215.

Fortin, C., Duchet, M. L., & Rousseau, R. (1996). Tapping sensitivity to processing in short-term memory. *Canadian Journal of Experimental Psychology, 50*, 402–407.

Fortin, C., & Rousseau, R. (1987). Time estimation as an index of processing demand in memory search. *Perception and Psychophysics, 42*, 377–382.

Fortin, C., & Rousseau, R. (1998). Interference from shorter-term memory processing on encoding and reproducing brief durations. *Psychological Research, 61*, 269–276.

Fortin, C., Rousseau, R., Bourque, P., & Kirouac, E. (1993). Time estimation and concurrent nontemporal processing: Specific interference from short-term memory demands. *Perception and Psychophysics, 53*, 536–548.

François, M. (1927). Contributions à l'étude du sens du temps: La température interne comme facteur de variation de l'appréciation subjective des durées. *L'Année Psychologique, 27*, 186–204.

Frankenhauser, M. (1959). *Estimation of time, an experimental study*. Stockholm: Almqvist and Wiksell.

Fraisse, P. (1967). *Psychologie du temps*. Paris: Presses Universitaires de France.

Franssen, V., & Vandierendonck, A. (2002). Time estimation: Does the reference memory mediate the effect of knowledge of results? *Acta Psychologica, 109*, 239–267.

Friedman, W. J. (Ed.). (1982). *The developmental psychology of time*. New York: Academic Press.

Friedman, W. J., & Janssen, S. M. J. (2010). Aging and the speed of time. *Acta Psychologica, 134*, 130–141.

Gautier, T., & Droit-Volet, S. (2002a). Attentional distraction and time perception in children. *International Journal of Psychology, 37*, 27–34.

Gautier, T., & Droit-Volet, S. (2002b). Attention and time estimation in 5- and 8-year-old children: A dual-task procedure. *Behavioural Processes, 58*, 57–66.

Geach, P. T. (1979). *An introduction to McTaggert's philosophy*. London: Hutchinson.

Gibbon, J. (1977). Scalar expectancy theory and Weber's Law in animal timing. *Psychological Review, 84*, 279–324.

Gibbon, J. (1979). Timing the stimulus and the response in aversive control. In M. D. Zeiler & P. Harzem (Eds.), *Reinforcement and the organisation of behaviour* (pp. 299–340). Chichester: Wiley.

Gibbon, J. (1981). On the form and location of the psychometric bisection function for time. *Journal of Mathematical Psychology, 24*, 58–87.

Gibbon, J. (1991). Origins of scalar timing. *Learning and Motivation, 22*, 3–38.

Gibbon, J., & Church, R. M. (1990). Representation of time. *Cognition, 37*, 23–54.

Gibbon, J., Church, R. M., & Meck, W. (1984). Scalar timing in memory. In J. Gibbon & L. Allan (Eds.), *Annals of the New York Academy of Sciences* (Timing and time perception, Vol. 423, pp. 52–77). New York: New York Academy of Sciences.

Gil, S., & Droit-Volet, S. (2009). Time perception, depression, and sadness. *Behavioural Processes, 80*, 169–176.

Gil, S., & Droit-Volet, S. (2011). "Time flies in the presence of angry faces"… depending on the temporal task used! *Acta Psychologica, 136*, 354–362.

Gil, S., & Droit-Volet, S. (2012). Emotional time distortions: The fundamental role of arousal. *Cognition and Emotion, 26*, 847–862.

Gil, S., Niedenthal, P. M., & Droit-Volet, S. (2007). Anger and time perception in children. *Emotion, 7*, 219–225.

Gil, S., Rousset, S., & Droit-Volet, S. (2009). How liked and disliked foods affect time perception. *Emotion, 9*, 457–463.

Goldstone, S., & Lhamon, W. T. (1974). Studies of auditory-visual differences in human time judgment: 1. Sounds are judged longer than lights. *Perceptual and Motor Skills, 39*, 63–82.

Gouvea, T. S., Monteiro, T., Soares, S., Atallah, B. V., & Paton, J. J. (2014). Ongoing behaviour predicts perceptual report interval duration. *Frontiers in Neurorobotics, 8*, 10.

Green, T. R. G., & Simpson, A. J. (1977). Time and temperature: A note on Bell. *Quarterly Journal of Experimental Psychology, 29*, 337–340.

Grommet, E. K., Droit-Volet, S., Gil, S., Hemmes, N. S., Baker, A. H., & Brown, B. L. (2011). Time estimation of fear cues in human observers. *Behavioural Processes, 86*, 88–93.

Gu, B.-M., & Meck, W. H. (2011). New perspectives on Vierordt's Law: Memory-mixing in ordinal temporal comparison tasks. In A. Vatakis, A. Esposito, M. Giagkou, F. A. Cummins, & G. Papadelis (Eds.), *Multidisciplinary aspects of time and time perception* (pp. 67–78). Berlin: Springer.

Guyau, M. (1890). *La genèse de l'idée de temps*. Paris: Alcan.

Haight, P. A., & Killeen, P. R. (1991). Adjunctive behaviour in multiple schedules of reinforcement. *Animal Learning and Behavior, 19*, 257–263.

Hammond, C. (2012). *Time warped: Unlocking the mysteries of time perception.* Edinburgh: Canongate.

Hedden, T., & Gabrieli, J. D. E. (2004). Insights into the ageing mind: A view from cognitive neuroscience. *Nature Reviews Neuroscience, 5*, 87–96.

Hellstrom, A. (1985). The time-order error and its relatives: Mirrors of cognitive comparing. *Psychological Bulletin, 97*, 35–61.

Helson, H. (1964). *Adaptation-level theory*. New York: Harper and Row.

Herbst, S. K., Javadi, A. H., van der Meer, E., & Busch, N. A. (2013). How long depends on how fast—perceived flicker dilates subjective duration. *PLoS ONE, 8*(10), e76074.

Hicks, R. E. (1992). Prospective and retrospective judgements of time: A neurobehavioral analysis. In F. Macar, V. Pouthas, & W. J. Friedman (Eds.), *Time, action, and cognition* (pp. 97–108). Dordrecht: Kluwer.

Hicks, R. E., Miller, G. W., & Kinsbourne, M. (1976). Prospective and retrospective judgments of time as a function of amount of information processed. *American Journal of Psychology, 89*, 719–730.

Hintzman, D. L., & Block, R. A. (1971). Repetition and memory: Evidence for a multiple-trace hypothesis. *Journal of Experimental Psychology, 88*, 297–306.

Hancock, P. A. (1993). Body temperature influence on time perception. *Journal of General Psychology*, 120, 197–216.

Hoagland, H. (1933). The physiological control of judgments of duration: Evidence for a chemical clock. *Journal of General Psychology, 9*, 267–287.

Hoagland, H. (1935). *Pacemakers in relation to aspects of behavior*. New York: MacMillan.

Hoagland, H. (1951). Consciousness and the chemistry of time. In H. Abramson (Ed.), *Problems of consciousness* (pp. 164–200). New York, Macy.

Hollingworth, H. L. (1910). The central tendency of judgment. *Journal of Philosophy, Psychology, and Scientific Methods, 7*, 461–469.

Ivry, R. B., & Hazeltine, R. E. (1992). Models of timing-with-a-timer. In F. Macar, V. Pouthas, & W. J. Friedman (Eds.), *Time, action, and cognition* (pp. 183–189). Dordrecht: Kluwer.

James, W. (1890). *The principles of psychology*. New York: H. Holt.

Jasselette, P., Lejeune, H., & Wearden, J. H. (1990). The perching response and the laws of animal timing. *Journal of Experimental Psychology: Animal Behavior Processes, 16*, 150–161.

Jazayeri, M., & Shadlen, M. N. (2010). Temporal context calibrates interval timing. *Nature Neuroscience, 13*, 1021–1028.

Jones, L. A., & Wearden, J. H. (2003). More is not necessarily better: Examining the nature of the temporal reference memory component in timing. *Quarterly Journal of Experimental Psychology, 56*, 321–343.

Jones, L. A., & Wearden, J. H. (2004). Double standards: Memory loading in temporal reference memory. *Quarterly Journal of Experimental Psychology, 57B*, 55–77.

Joubert, C. E. (1990). Subjective expectations of the acceleration of time with aging. *Perceptual and Motor Skills, 70*, 334.

Jozefowiez, J., Polack, C. W., Machado, A., & Miller, R. R. (2014). Trial frequency effects in human temporal bisection: Implications for theories of timing. *Behavioural Processes, 101*, 81–88.

Jozefowiez, J., Staddon, J. E. R., & Cerutti, D. T. (2009). The behavioural economics of choice and interval timing. *Psychological Review, 116*, 519–539.

Kant, I. (1781/1900/2003). *Critique of pure reason* (J. M. D. Meiklejohn, Trans.). Mineola, NY: Dover Publications.

Killeen, P. R., & Fetterman, J. G. (1988). A behavioral theory of timing. *Psychological Review, 95*, 274–295.

Klapproth, F., & Müller, M. (2008). Temporal generalization under time pressure in humans. *Quarterly Journal of Experimental Psychology, 61*, 588–600.

Klapproth, F., & Wearden, J. H. (2011). Why do temporal generalization gradients change when people make decisions as quickly as possible? *Quarterly Journal of Experimental Psychology, 64*, 1646–1664.

Kopec, C. D., & Brody, C. D. (2010). Human performance on the temporal bisection task. *Brain and Cognition, 74*, 262–272.

Kramer, G., & Rilling, G. M. (1970). Differential reinforcement of low rates: A selective critique. *Psychological Bulletin, 74*, 225–254.

Kuhn, G., & Tipples, J. (2011). Increased gaze following for fearful faces. It depends on what you're looking for! *Psychonomic Bulletin and Review, 18*, 89–95.

Lambrechts, A., Mella, N., Pouthas, V., & Noulhiane, M. (2011). Subjectivity of time perception: A visual emotional orchestration. *Frontiers in Integrative Neuroscience, 5*, 73.

Lang, P. J., Bradley, M. M., & Cuthbert, B. N. (2008). *International affective picture system (IAPS): Affective ratings of pictures and instruction manual* (Tech. Rep. A-8). Gainesville, FL: University of Florida.

Langer, J., Wapner, S., & Werner, H. (1961). The effect of danger upon the experience of time. *American Journal of Psychology, 74*, 94–97.

Larson, E. (2004). The time of our lives: The experience of temporality in occupation. *Canadian Journal of Occupational Therapy, 1*, 22–32.

Larson, E., & von Eye, A. (2006). Predicting the perceived flow of time from qualities of activity and depth of engagement. *Ecological Psychology, 18*, 113–130.

Leak, T. M., & Gibbon, J. (1995). Simultaneous timing of multiple intervals: Implications for the scalar property. *Journal of Experimental Psychology: Animal Behavior Processes*, 2, 3–19.

Lejeune, H. (1998). Switching or gating? The attentional challenge in cognitive models of psychological time. *Behavioural Processes, 44*, 145

Lejeune, H., Cornet, S., Ferreira, J., & Wearden, J. H. (1998). How do Mongolian gerbils (Meriones unguiculatus) pass the time? Adjunctive behavior during temporal differentiation in gerbils. *Journal of Experimental Psychology: Animal Behavior Processes, 24*, 325–334.

Lejeune, H., & Jasselette, P. (1986). Accurate DRL performance in the pigeon: Comparison between perching and treadle pressing. *Animal Learning and Behavior, 14*, 205–211.

Lejeune, H., Richelle, M., & Wearden, J. H. (2006). About Skinner and time: Behavior-analytic contributions to research on animal timing. *Journal of the Experimental Analysis of Behavior, 85*, 125–142.

Lejeune, H., & Wearden, J. H. (1991). The comparative psychology of fixed-interval responding: Some quantitative analyses. *Learning and Motivation, 22*, 84–111.

Lejeune, H., & Wearden, J. H. (2006). Scalar properties in animal timing: Conformity and violations. *Quarterly Journal of Experimental Psychology, 59*, 1875–1908.

Lejeune, H., & Wearden, J. H. (2009). Vierordt's "The experimental study of the time sense" (1868) and its legacy. *European Journal of Cognitive Psychology, 21*, 941–960.

Lemlich, R. (1975). Subjective acceleration of time with aging. *Perceptual and Motor Skills, 41*, 235–238.

Levin, I. (1977). The development of time concepts in young children: Reasoning about duration. *Child Development, 48*, 435–444.

Levin, I. (1979). Interference of time-related and unrelated cues with duration comparisons of young children: Analysis of Piaget's formulation of the relation of time and speed. *Child Development, 50*, 469–477.

Levin, I. (1992). The development of the concept of time in children: An integrative model. In F. Macar, V. Pouthas, & W. J. Friedman (Eds.), *Time, action, and cognition* (pp. 13–32). Dordrecht: Kluwer.

Lowe, C. F., Harzem, P., & Spencer, P. T. (1979). Temporal control of behavior and the power law. *Journal of the Experimental Analysis of Behavior, 31*, 333–344.

Lustig, C., & Meck, W. H. (2001). Paying attention to time as one gets older. *Psychological Science, 12*, 478–484.

Macar, F., Grondin, S., & Casini, L. (1994). Controlled attention sharing influences time estimation. *Memory and Cognition, 22*, 673–686.

Machado, A. (1997). Learning the temporal dynamics of behavior. *Psychological Review, 104*, 241–265.

Machado, A., & Keen, R. (1999). Learning to Time (LET) or Scalar Expectancy Theory (SET)? A critical test of two models of timing. *Psychological Science, 10*, 285–290.

Machado, A., Malheiro, M. T., & Erlhagen, W. (2009). Learning to time: A perspective. *Journal of the Experimental Analysis of Behavior, 92*, 423–458.

Malapani, C., Deweer, B., & Gibbon, J. (2002). Separating storage from retrieval dysfunction of temporal memory in Parkinson 's disease. *Journal of Cognitive Neuroscience, 14*, 311–322.

Matell, M. S., & Meck, W. H. (2004). Cortico-striatal circuits and interval timing: coincidence detection of oscillatory processes. *Cognitive Brain Research, 21*, 139–170.

Mattes, S., & Ulrich, R. (1998). Directed attention prolongs the perceived duration of a brief stimulus. *Perception and Psychophysics, 60*, 1305–1317.

McAuley, J. D., Miller, J. P., Wang, M., & Pang, K. C. H. (2010). Dividing time: concurrent timing of auditory and visual events by young and elderly adults. *Experimental Aging Research, 36*, 306–324.

McClain, L. (1983). Interval estimation: Effects of processing demands on prospective and retrospective reports. *Perception and Psychophysics, 34*, 185–189.

McCormack, T., Brown, G. D. A., Maylor, E. A., Darby, R. J., & Green, D. (1999). Developmental changes in time estimation: Comparing childhood and old age. *Developmental Psychology, 35*, 1143–1155.

McCormack, T., Brown, G. D. A., Smith, M. C., & Brock, J. (2004). A timing specific memory distortion effect in young children. *Journal of Experimental Child Psychology*, 87, 33–56.

McCormack, T., Wearden, J. H., Smith, M. C., & Brown, D. G. A. (2005). Episodic temporal generalization: A developmental study. *Quarterly Journal of Experimental Psychology, 58A*, 693–704.

McTaggart, J. M. E. (1908). The unreality of time. *Mind, 17*, 457–473.

Meck, W. H. (1983). Selective adjustment of the speed of internal clock and memory processes. *Journal of Experimental Psychology: Animal Behavior Processes, 9*, 171–201.

Meissner, K., & Wittmann, M. (2011). Body signals, cardiac awareness, and the perception of time. *Biological Psychology, 86*, 289–297.

Mensch, J. (2014). A brief account of Husserl's conception of our consciousness in time. In V. Arstila & D. Lloyd (Eds.), *Subjective time: The philosophy, psychology, and neuroscience of temporality* (pp. 43–59). Cambridge, MA: MIT Press.

Miyake, A., Friedman, P., Emerson, M. J., Witzki, A. H., Howerter, A., & Wager, T. D. (2000). The unity and diversity of executive functions and their contributions to complex "frontal lobe" tasks: A latent variable analysis. *Cognitive Psychology, 41*, 49–100.

Monsell, S. (2003). Task switching. *Trends in Cognitive Science, 7*, 134–140.

Montangero, J. (1984). Perspectives actuelles sur la psychogenèse du temps. *L'Année Psychologique, 84*, 433–460.

Nichols, H. (1891). The psychology of time. *American Journal of Psychology, 3*, 453–529.

Nisbett, R. E., & Wilson, T. D. (1977). Telling more than we can know: Verbal reports on mental processes. *Psychological Review, 84*, 231–259.

Noulhiane, M., Melia, N., Samson, S., Ragot, R., & Pouthas, V. (2007). How emotional auditory stimuli modulate time perception. *Emotion, 7*, 697–704.

Noyes, R., & Kletti, R. (1976). Depersonalization in the face of life-threatening danger: A description. *Psychiatry, 39*, 19–27.

Ogden, R. S., & Jones, L. A. (2009). More is still not better: Testing the perturbation model of temporal reference memory across different modalities and tasks. *Quarterly Journal of Experimental Psychology, 62*, 909–924.

Ogden, R., Wearden, J. H., & Jones, L. A. (2008). The remembrance of times past: Interference in temporal reference memory. *Journal of Experimental Psychology: Human Perception and Performance, 34*, 1524–1544.

Ogden, R. S., Wearden, J. H., Gallagher, D. T., & Montgomery, C. (2011). The effect of alcohol administration on human timing: A comparison of prospective timing, retrospective timing, and passage of time judgements. *Acta Psychologica, 138*, 254–262.

Ogden, R. S., Wearden, J. H., & Montgomery, C. (2014). The differential contribution of executive functions to temporal generalization reproduction, and verbal estimation. *Acta Psychologica, 152*, 84–94.

Ono, F., & Kitazawa, S. (2011). Shortening of subjective visual intervals followed by repetitive stimulation. *PLoS ONE, 6*(12), e28722.

Ornstein, R. E. (1969). *On the experience of time*. London: Penguin.

Ortega, L., & Lopez, F. (2008). Effects of visual flicker on subjective time in a temporal bisection task. *Behavioural Processes, 78*, 380–386.

Paul, I., Wearden, J., Bannier, D., Gontier, E., Le Dantec, C., & Rebai, M. (2011). Making decisions about time: Event-related potentials and judgements about the equality of durations. *Biological Psychology, 88*, 94–103.

Pavlov, I. P. (1960). *Conditioned reflexes*. New York: Dover Publications. Original edition 1927.

Penton-Voak, I. S., Edwards, H., Percival, A., & Wearden, J. H. (1996). Speeding up an internal clock in humans? Effects of click trains on subjective duration. *Journal of Experimental Psychology: Animal Behavior Processes, 22*, 307–320.

Pfaff, D. (1968). Effects of temperature and time of day on time judgments. *Journal of Experimental Psychology, 76*, 419–422.

Phillips, I. (2013). Perceiving the passing of time. *Proceedings of the Aristotelean Society, 113*, 225–252.

Phillips, I. (2014). The temporal structure of experience. In V. Arstila & D. Lloyd (Eds.), *Subjective time: The philosophy, psychology, and neuroscience of temporality* (pp. 139–158). Cambridge, MA: MIT Press.

Piaget, J. (1969). *The child's conception of time*. London: Routledge and Kegan Paul.

Platt, J. R. (1979). Temporal differentiation and the psychophysics of time. In M. D. Zeiler & P. Harzem (Eds.), *Reinforcement and the organisation of behaviour* (pp. 1–29). Chichester: Wiley.

Pollatos, O., Laubrock, J., & Wittmann, M. (2014). Interoceptive focus shapes the experience of time. *PLoS ONE, 9*(1), e86934.

Predebon, J. (1996). The effect of active and passive processing of interval events on prospective and retrospective time estimates. *Acta Psychologica, 94*, 41–58.

Provasi, J., Rattat, A.-C., & Droit-Volet, S. (2010). Temporal bisection in 4-month-old infants. *Journal of Experimental Psychology: Animal Behavior Processes, 37*, 108–113.

Rakitin, B. C., Gibbon, J., Penney, T. B., Malapani, C., Hinton, S. C., & Meck, W. H. (1998). Scalar expectancy theory and peak-interval timing in humans. *Journal of Experimental Psychology: Animal Behavior Processes, 24*, 15–33.

Rammsayer, T. G., & Brandler, S. (2007). Performance on temporal information processing as an index of general intelligence. *Intelligence, 35*, 123–139.

Rattat, A.-C., & Droit-Volet, S. (2005). Development of long term memory for duration in a temporal bisection task. *Quarterly Journal of Experimental Psychology, 58B*, 163–176.

Rattat, A.-C., & Droit-Volet, S. (2007). Implicit long-term memory for duration in young children. *European Journal of Cognitive Psychology, 19*, 271–285.

Rescorla, R. A. (1991). Associative relations in instrumental learning: The eighteenth Bartlett Lecture. *Quarterly Journal of Experimental Psychology, 43B*, 1–23.

Reuter-Lorentz, P. A., & Lustig, C. (2005). Brain aging: Reorganizing discoveries about the aging mind. *Current Opinion in Neurobiology, 15*, 245–251.

Richards, D. D. (1982). Children's time concepts: Going the distance. In W. J. Friedman (Ed.), *The developmental psychology of time* (pp. 13–45). New York: Academic Press.

Richelle, M., & Lejeune, H. (1980). *Time in animal behaviour.* Oxford: Pergamon.

Roberts, S. (1981). Isolation of an internal clock. *Journal of Experimental Psychology: Animal Behavior Processes, 7*, 242–268.

Ryan, L. J., & Fritz, M. S. (2007). Erroneous knowledge of results affects decision and memory processes on timing tasks. *Journal of Experimental Psychology: Human Perception and Performance, 33*, 1468–1482.

Skinner, B. F. (1938). *The behavior of organisms.* New York: Appleton-Century-Crofts.

Skinner, B. F. (1948). "Superstition" in the pigeon. *Journal of Experimental Psychology, 38*, 168–172.

Spetch, M. L., & Wilkie, D. M. (1983). Subjective shortening: A model of pigeons' memory for event duration. *Journal of Experimental Psychology: Animal Behavior Processes, 9*, 14–30.

Staddon, J. E. R., & Ayres, S. (1975). Sequential and temporal properties of behaviour induced by a schedule of periodic food delivery. *Behaviour, 54*, 26–49.

Staddon, J. E. R., & Higa, J. (1999). Time and memory: Towards a pacemaker-free theory of interval timing. *Journal of the Experimental Analysis of Behavior, 71*, 215–251.

Staddon, J. E. R., & Simmelhag, V. L. (1971). The "superstition" experiment: A re-examination of its implications for the principles of adaptive behavior. *Psychological Review, 78*, 3–43.

Sternberg, S. (1966). High-speed scanning in human memory. *Science, 153*, 652–654.

Stetson, C., Fiesta, M. P., & Eagleman, D. M. (2007). Does time really slow down during a frightening event? *PLOS One, 2*(12), e1295.

Taatgen, N., van Rijn, H., & Anderson, J. (2007). An integrated theory of prospective time interval estimation: The role of cognition, attention, and learning. *Psychological Review, 114*, 577–598.

Takahashi, K., & Watanabe, K. (2012). Short-term memory for event duration: Modality specificity and goal dependency. *Attention, Perception, and Psychophysics, 74*, 1623–1631.

Thayer, R. E. (1967). Measurement of activation through self-report. *Psychological Reports, 20*, 663–678.

Thomas, E. A. C., & Weaver, W. B. (1975). Cognitive processing and time perception. *Perception and Psychophysics, 17*, 363–367.

Tillman, K. A., & Barner, D. (2015). Learning the language of time: Children's acquisition of duration words. *Cognitive Psychology, 78*, 57–77.

Tipples, J. (2008). Negative emotionality influences the effects of emotion on time perception. *Emotion, 8*, 127–131.

Tipples, J. (2010). Time flies when we read taboo words. *Psychonomic Bulletin and Review, 17*, 563–568.

Treisman, M. (1963). Temporal discrimination and the indifference interval: Implications for a model of the "internal clock". *Psychological Monographs, 77*, whole number 576.

Treisman, M. (2013). The information-processing model of timing (Treisman, 1963): Its sources and further development. *Timing and Time Perception, 1*, 131–158.

Treisman, M., Faulkner, A., Naish, P. L. N., & Brogan, D. (1990). The internal clock: Evidence for a temporal oscillator underlying time perception with some estimates of its characteristic frequency. *Perception, 19*, 705–748.

Troche, S. J., & Rammsayer, T. H. (2009). The influence of temporal resolution power and working memory capacity on psychometric intelligence. *Intelligence, 37*, 479–486.

Troche, S. J., Wagner, F. L., Voelke, A. E., Roebers, C. M., & Rammsayer, T. H. (2014). Individual differences in working memory capacity explain the rela-

tionship between general discrimination ability and psychometric intelligence. *Intelligence, 44*, 40–50.
Tse, P. U., Intriligator, J., Rivest, J., & Cavanagh, P. (2004). Attention and the subjective experience of time. *Perception and Psychophysics, 66*, 1171–1189.
Underwood, B. J. (1969). Attributes of memory. *Psychological Review, 76*, 559–573.
Vanneste, S., Pouthas, V., & Wearden, J. H. (2001). Temporal control of rhythmic performance: A comparison between young and old adults. *Experimental Aging Research, 27*, 83–102.
Vierordt, K. (1868). *Der Zeitsinn nach Versuchen*. Tubingen: Laupp.
Vroon, P. A. (1970). Effect of presented and processed information on duration experience. *Acta Psychologica, 34*, 115–121.
Wackerman, J., & Ehm, W. (2006). The dual klepsydra model of time representation and time reproduction. *Journal of Theoretical Biology, 239*, 482–493.
Wearden, J. H. (1985). The power law and Weber's law in fixed-interval postreinforcement pausing: A scalar timing model. *Quarterly Journal of Experimental Psychology, 37B*, 191–211.
Wearden, J. H. (1990). Maximizing reinforcement rate on spaced responding schedules under conditions of temporal uncertainty. *Behavioural Processes, 22*, 47–59.
Wearden, J. H. (1991a). Do humans possess an internal clock with scalar timing properties? *Learning and Motivation, 22*, 59–83.
Wearden, J. H. (1991b). Human performance on an analogue of an interval bisection task. *Quarterly Journal of Experimental Psychology, 43B*, 59–81.
Wearden, J. H. (1992). Temporal generalization in humans. *Journal of Experimental Psychology: Animal Behavior Processes, 18*, 134–144.
Wearden, J. H. (1995). Categorical scaling of stimulus duration by humans. *Journal of Experimental Psychology: Animal Behavior Processes, 21*, 318–330.
Wearden, Denovan, Fahkri, and Haworth (1997)
Wearden, J. H. (1999). "Beyond the fields we know…": Exploring and developing scalar timing theory. *Behavioural Processes, 45*, 3–21.
Wearden, J. H. (2002). Traveling in time: A time-left analogue for humans. *Journal of Experimental Psychology: Animal Behavior Processes, 28*, 200–208.
Wearden, J. H. (2003). Applying the scalar timing model to human time psychology: Progress and challenges. In H. Helfrich (Ed.), *Time and mind II: Information-processing perspectives* (pp. 21–39). Gottingen: Hogrefe & Huber.
Wearden, J. H. (2004). Decision processes in models of timing. *Acta Neurobiologiae Experimentalis, 64*, 303–317.

Wearden, J. H. (2005). The wrong tree: Time perception and time experience in the elderly. In J. Duncan, L. Phillips, & P. McLeod (Eds.), *Measuring the mind: Speed, age, and control* (pp. 137–158). Oxford: Oxford University Press.

Wearden, J. H. (2008a). Slowing down an internal clock: Implications for accounts of performance on four timing tasks. *Quarterly Journal of Experimental Psychology, 61*, 263–274.

Wearden, J. H. (2008b). The perception of time: Basic research and some potential links to the study of language. *Language Learning, 58*(Suppl. 1), 149–171.

Wearden, J. H. (2013a). "From that paradise": The golden anniversary of timing. *Timing and Time Perception, 1*, 127–130.

Wearden, J. H. (2013b). The cognitive neuroscience of time perception: How psychological studies might help dissect the timing system. *Neuropsychologia, 51*, 187–190.

Wearden, J. H. (2015). Passage of time judgements. *Consciousness and Cognition, 38*, 165–171.

Wearden, J. H., & Bray, S. (2001). Scalar timing without reference memory: Episodic temporal generalization and bisection in humans. *Quarterly Journal of Experimental Psychology, 54B*, 289–310.

Wearden, J. H., Denovan, L., Fakhri, M., & Haworth, R. (1997). Scalar timing in temporal generalization in humans with longer stimulus durations. *Journal of Experimental Psychology: Animal Behavior Processes, 23*, 502–511.

Wearden, J. H., Edwards, H., Fakhri, M., & Percival, A. (1998). Why "sounds are judged longer than lights": Application of a model of the internal clock in humans. *Quarterly Journal of Experimental Psychology, 51B*, 97–120.

Wearden, J. H., & Farrar, R. (2007). Effects of feedback and calibration on the verbal estimation of the duration of tones. *Acta Psychologica, 126*, 1–17.

Wearden, J. H., & Ferrara, A. (1993). Subjective shortening in humans' memory for stimulus duration. *Quarterly Journal of Experimental Psychology, 46B*, 163–186.

Wearden, J. H., & Ferrara, A. (1995). Stimulus spacing effects in temporal bisection by humans. *Quarterly Journal of Experimental Psychology, 48B*, 289–310.

Wearden, J. H., Goodson, G., & Foran, K. (2007). Subjective shortening with filled and unfilled auditory and visual intervals in humans? *Quarterly Journal of Experimental Psychology, 60*, 1616–1628.

Wearden, J. H., & Grindrod, R. (2003). Manipulating decision processes in the human scalar timing system. *Behavioural Processes, 61*, 47–56.

Wearden, J. H., & Jones, L. A. (2007). Is the growth of subjective time in humans a linear or non-linear function of real time? *Quarterly Journal of Experimental Psychology, 60*, 1289–1302.

Wearden, J. H., & Jones, L. A. (2013). Explaining between-group differences in performance on timing tasks. *Quarterly Journal of Experimental Psychology, 66*, 179–199.

Wearden, J. H., & Lejeune, H. (2006). "The stone which the builders rejected": Delay of reinforcement and response rate on fixed-interval and related schedules. *Behavioural Processes, 71*, 77–87.

Wearden, J. H., & Lejeune, H. (2008). Scalar properties in human timing: Conformity and violations. *Quarterly Journal of Experimental Psychology, 61*, 569–587.

Wearden, J. H., & McShane, B. (1988). Interval production as an analogue of the peak procedure: Evidence for similarity of human and animal timing processes. *Quarterly Journal of Experimental Psychology, 40B*, 363–375.

Wearden, J. H., Norton, R., Martin, S., & Montford-Bebb, O. (2007). Internal clock processes and the filled duration illusion. *Journal of Experimental Psychology: Human Perception and Performance, 33*, 716–729.

Wearden, J. H., O'Donoghue, A., Ogden, R., & Montgomery, C. (2014). Subjective duration in the laboratory and the world outside. In V. Arstila & D. Lloyd (Eds.), *Subjective time: The philosophy, psychology, and neuroscience of temporality* (pp. 287–306). Cambridge, MA: MIT Press.

Wearden, J. H., O'Rourke, S. C., Matchwick, C., Zhang, M., & Maeers, S. (2010). Task switching and subjective duration. *Quarterly Journal of Experimental Psychology, 63*, 531–543.

Wearden, J. H., Parry, A., & Stamp, L. (2002). Is subjective shortening in human memory unique to time representations? *Quarterly Journal of Experimental Psychology, 55B*, 1–25.

Wearden, J. H., & Penton-Voak, I. S. (1995). Feeling the heat: Body temperature and the rate of subjective time, revisited. *Quarterly Journal of Experimental Psychology, 48B*, 129–141.

Wearden, J. H., Smith-Spark, J. H., Cousins, R., Edelstyn, N. M. J., Cody, F. W. J., & O'Boyle, D. J. (2008). Stimulus timing by people with Parkinson's disease. *Brain and Cognition, 67*, 264–279.

Wearden, J. H., Smith-Spark, J. H., Cousins, R., Edelstyn, N. M. J., Cody, F. W. J., & O'Boyle, D. J. (2009). Effect of click trains on duration estimates by people with Parkinson's disease. *Quarterly Journal of Experimental Psychology, 62*, 33–40.

Wearden, J. H., Todd, N. P. M., & Jones, L. A. (2006). When do auditory/visual differences in duration judgements occur? *Quarterly Journal of Experimental Psychology, 59*, 1709–1724.

Wearden, J. H., Wearden, A. J., & Rabbitt, P. (1997). Age and IQ effects on stimulus and response timing. *Journal of Experimental Psychology: Human Perception and Performance, 23*, 962–979.

Wechsler, D. (1998). *WAIS III and WMS-III manual.* San Antonio, TX: Psychological Corporation.

Whitaker, J. S., Lowe, C. F., & Wearden, J. H. (2003). Multiple-interval timing in rats: Performance on two-valued mixed fixed-interval schedules. *Journal of Experimental Psychology: Animal Behavior Processes, 29*, 277–291.

Whitaker, S., Lowe, C. F., & Wearden, J. H. (2008). When to respond? And how much? Temporal control and response output on mixed-fixed-interval schedules with unequally-probable components. *Behavioural Processes, 77*, 33–42.

Wing, A. M., & Kristofferson, A. B. (1973). Response delays and the timing of discrete motor responses. *Perception and Psychophysics, 14*, 3–12.

Wittman, M., & Lehnhoff, S. (2005). Age effects in the perception of time. *Psychological Reports, 97*, 921–935.

Woodrow, H. (1930). The reproduction of temporal intervals. *Journal of Experimental Psychology, 13*, 473–499.

Zakay, D. (1992). The role of attention in children's time perception. *Journal of Experimental Child Psychology, 54*, 355–371.

Zakay, D. (1993). Relative and absolute duration judgments under prospective and retrospective paradigms. *Perception and Psychophysics, 54*, 656–664.

Zakay, D., & Block, R. A. (1995). An attentional gate model of prospective time estimation. In M. Richelle, V. DeKeyser, G. d'Ydewalle, & A. Vandierendock (Eds.), *Time and the dynamic control of behavior* (pp. 167–178). Liège: Université de Liège, P.A.I.

Zélanti, P. S., & Droit-Volet, S. (2011). Cognitive abilities explaining age-related changes in time perception of short and long durations. *Journal of Experimental Child Psychology, 109*, 143–157.

Zélanti, P. S., & Droit-Volet, S. (2012). Auditory and visual differences in time perception? An investigation from a developmental perspective with neuropsychological tests. *Journal of Experimental Child Psychology, 112*, 296–311.

Author Index

J. Wearden, *The Psychology of Time Perception*,
DOI 10.1057/978-1-137-40883-9

L

M

N

Subject Index

J. Wearden, *The Psychology of Time Perception*,
DOI 10.1057/978-1-137-40883-9

www.ingramcontent.com/pod-product-compliance
Lightning Source LLC
Chambersburg PA
CBHW050811180926
51622CB00043B/106
* 9 7 8 1 1 3 7 4 0 8 8 2 2 *